Certified Kubernetes Administrator (CKA) Exam Guide

Validate your knowledge of Kubernetes and implement it
in a real-life production environment

Mélony Qin

BIRMINGHAM—MUMBAI

Certified Kubernetes Administrator (CKA) Exam Guide

Group Product Manager: Rahul Nair

Publishing Product Manager: Niranjan Naikwadi

Senior Editor: Arun Nadar

Content Development Editor: Sujata Tripathi

Technical Editor: Arjun Varma

Copy Editor: Safis Editing

Project Coordinator: Ashwin Dinesh Kharwa

Proofreader: Safis Editing

Indexer: Sejal Dsilva

Production Designer: Vijay Kamble

Marketing Coordinator: Nimisha Dua

First published: November 2022

Production reference: 1071022

Published by Packt Publishing Ltd.

Livery Place

35 Livery Street

Birmingham

B3 2PB, UK.

ISBN 978-1-80323-826-5

www.packt.com

Foreword

Over the last decade, Kubernetes has gone mainstream. Builders of cloud applications are expected to be familiar with cloud-native design tools and techniques. Becoming certified in Kubernetes demonstrates that you have the knowledge and skills necessary to meet the expectations of businesses, enterprises, and consumers.

Whether you are a cloud-native expert or a beginner, this book will familiarize you with the tools, technologies, and terminology in the cloud-native ecosystem. Mélony's own experience getting involved in the cloud-native ecosystem and its rapidly changing array of open source projects and cloud-based products, enable her to write an approachable book that can serve as your guide to the modern way that today's applications are built.

The expectations of business, customers, and users of today's applications have never been greater. Kubernetes and cloud-native are the skills that will enable you to build applications that meet the standards necessary to compete in the world of modern application development.

Brendan Burns, co-founder of Kubernetes open source project.

I first met Mélony back in 2018. I'd been speaking at an event in London and when I finished, she approached the stage to ask a question. That was just a few days before she started working at Microsoft and I've had the pleasure of working with her since then. It's rare to get to work with such a talented individual, someone who has a real passion for technology, learning, and helping others to learn.

I've been working in technology for over 30 years in roles across operations, engineering, and architecture. A lot of that time was spent working in large corporations. Containers and Kubernetes have had a massive effect on the way applications are developed, deployed, and managed. It would have solved so many problems if I'd had these tools available earlier on in my career. Back in 2014, a colleague of mine told me to keep a close eye on this "Docker" thing they'd heard about as they were convinced it was going to be a big deal. I have kept an eye on it and they were right. It has been a big deal!

As your maturity with containers grows, you'll find yourself involved with Kubernetes. If you're going to use Kubernetes, then you really need to understand how it works. It's complicated and you can easily get things wrong, so you really, really need to know what you're doing. Certification exams are always a great way to build your knowledge, test yourself, and prove that you know your stuff! I've sat all of the current Kubernetes certification exams and I can tell you from personal experience, these exams are hard. You can't get away with guessing which multiple-choice answer is the right one. You can't wing it. Oh no, you need to actually put the work in to learn Kubernetes before you sit this exam! And that's a good thing, because it makes the Kubernetes certification more valuable knowing that you can't pass it without putting in the effort.

If you've got this book in your hands or on your digital reading device of choice, then you've made a great start! Kubernetes certifications are hard, make no mistake, but you will pass if you put the work in. Mélony will guide you through the topics you need to learn and help set you up for success.

Good luck. You've got this!

Mark Whitby, Cloud-native architecture and engineering lead, principal global black belt (GBB) at Microsoft

Certifications are the best way to show the world your passion, your interests, and your skills, in the ever competitive and fierce landscape for talent sprung by the adoption of the cloud native paradigm. Mélony has done an amazing job to demystify the many mysteries of Kubernetes into simple, easy to understand concepts that will guide you in your studies, and hopefully lead you to a successful certification. She truly understands the learning journey and the many hurdles of cloud native, and she's motivated to make your journey easier.

I met Mélony several years ago at one of the Microsoft OpenHacks events, and her passion for learning and sharing struck me. So, I'm both glad and proud to write the foreword to her new book, which undoubtedly will help many cloud native engineers in their own personal learning path.

Alessandro Vozza, Principal Software Engineer at Microsoft, CNCF Ambassador, Founder of Cloud Pirates

Note from the author

Containerization is an approach to managing applications; a container image contains all its deployment dependencies and configurations. Managing one, or even a couple of containers for dev/testing purposes, is relatively easy. The real challenge comes when you have to manage hundreds, or even thousands of containers, especially for enterprise-grade product environments, where you'll be managing networking, deployments, configuration, etc. This is where the container orchestrator comes in.

Looking back, many open source container orchestrators have been popular in the market at one point in time. Although we're still hearing about Docker Swarm, Mesosphere's DC/OS, Kubernetes is by far the most popular container orchestration tool.

We have seen tremendous growth in Kubernetes and its ecosystem over the last 7 years. Yet, the complexity of managing the tool remains the major blocker for enterprises that prevents them from taking complete advantage of this fantastic technology. Learning Kubernetes and its ecosystem will help organizations overcome their challenges in deploying, managing, and operating Kubernetes clusters.

Acquiring a **Certified Kubernetes Administrator (CKA)** certification is the best way to help you train the essential skills on working with Kubernetes. In particular, you'll learn how to manage and operate Kubernetes.

The **Certified Kubernetes Administrator (CKA)** certification is founded by **Cloud Native Computing Foundation (CNCF)**, and it is designed to ensure that certification candidates have the skills and knowledge to help them establish their credibility and value in the job market, and to support business growth. It is widely recognized by various sizes of businesses across different industries.

This book is an exam guide and a knowledge book, and it covers all the important aspects required by the CKA certification. We'll start with an introduction to Kubernetes architecture, turning to the core concept of Kubernetes. Then, we will dive deeply into the main Kubernetes primitives, installation and configuration, cluster management, workload scheduling, networking, and security. We'll also cover various ways to troubleshoot Kubernetes.

Each chapter will cover core concepts as well as code samples. It is not a book to read conventionally – it is a practice guide that requires you to get out of your comfort zone and go break some eggs!

While I was writing this book, I was at the lowest point of my life, having relocated to a new continent, as well as undergoing surgery for the first time in my life during the first 2 months of relocation. This all took place alongside many other challenges. I can't thank my family enough for the huge support I received from them, especially my beloved mother, Nancy Deng. I also want to thank my lovely local and remote friends, the Packt team, and other people who supported me during that period.

As a human being, those unprecedented life challenges also made me rethink the definition of living a meaningful life. Hence, I decided to turn those challenges into something positive and meaningful by pushing myself to the max to work on this book. This experience also encouraged me to create the CloudMelon Vis YouTube channel, alongside my website `cloud-melon.com` that I have been blogging on for years. Sharing is caring!

Rethinking my community evangelization in the past, I hope to make my life more meaningful by making a more positive impact on the community. This book aims to help people find their new career path with Kubernetes, in particular those who lost their jobs during the pandemic. Kubernetes is one of the most life-changing technologies that empowered my own career path, and I hope it will make a positive impact on your career, too.

Last but not least, I wish you the best of luck with your CKA exam and hope you will enjoy your journey in building your future with this book. Thanks!

Contributors

About the author

Mélony Qin, aka CloudMelon, is the founder of CloudMelonVision and a product manager at a top tech company, as well as being the author of *Microsoft Azure Infrastructure*, the *Kubernetes Workshop*, and *Certified Kubernetes Administrator (CKA) Exam Guide* by Packt Publishing, and the technical reviewer for *Azure for Architects, Third Edition*. Her community contribution mainly concerns OSS, DevOps, Kubernetes, serverless, big data analytics, and IoT on Microsoft Azure. She is also a member of the **Association for Computing Machinery (ACM)** and **Project Management Institute (PMI)**. She can be reached via Twitter using @MelonyQ or @CloudMelonVis, through the Contact me page of her blog (www.cloud-melon.com), and via her YouTube channel: CloudMelon Vis https://www.youtube.com/c/CloudMelonVis.

About the reviewers

Erol Kavas has worked in the IT industry for more than 20 years, with 10 years dedicated to infrastructure, the cloud, and DevOps. He has helped many Canadian and US enterprises and governments to build their cloud foundations and embark upon their containerization and Kubernetes journeys. He is fully certified on AWS, Azure, Google Cloud Platform, and Kubernetes in all disciplines. He is a partner and chief consultant in a DevOps and cloud consulting firm that helps Canadian and US start-ups in their cloud and DevOps journeys. He is also a **Microsoft Certified Trainer** (**MCT**) regional lead for Canada and trains many new cloud professionals at CloudCamp.ca.

Dustin Specker has been in the tech industry for almost 10 years. He started as a frontend web developer focused on usability. In the last few years, Dustin has pivoted to developing cloud solutions. He has used Kubernetes for on-premises environments and public cloud for the last four years. He has earned the CKAD, CKA, and CKS certifications. He received a Bachelor of Science degree in nuclear engineering from the Missouri University of Science and Technology, where he discovered that he enjoyed programming much more than nuclear engineering.

Bruno S. Brasil is a cloud engineer who has used Linux since he was a kid. He started out working in on-premises environments before living out the migration to cloud solutions and joining the DevOps culture, choosing Google Cloud Platform as his specialization focus. Since then, he has worked on projects of this type as a consultant and engineer for several types of businesses, ranging from digital banks and marketplaces to start-ups. He has always focused on implementing best practices in the development of infrastructure as code, disseminating the DevOps culture, and implementing SRE strategies. He is enthusiastic about the open source community and believes that this is the most important path in terms of the growth of new professionals and new technologies.

Juri Sinar is a senior DevOps engineer working for a London fintech start-up. Kubernetes is the main platform that he has used to run and integrate infrastructure for the past five years. It helps Juri to connect and automate a large global network of open banking for his clients in a way that would not have been possible just 10 years ago.

Table of Contents

Part 1: Cluster Architecture, Installation, and Configuration

1

Kubernetes Overview   3

2

Installing and Configuring Kubernetes Clusters   21

3

Maintaining Kubernetes Clusters   47

Part 2: Managing Kubernetes

4

Application Scheduling and Lifecycle Management   71

5

Demystifying Kubernetes Storage   117

6

Securing Kubernetes   143

7

Demystifying Kubernetes Networking   165

Part 3: Troubleshooting

8

Monitoring and Logging Kubernetes Clusters and Applications   207

9

Troubleshooting Cluster Components and Applications   227

10

Troubleshooting Security and Networking   253

Appendix - Mock CKA scenario-based practice test resolutions   267

Preface

Kubernetes is by far the most popular container orchestration tool, yet the complexities of managing the tool have led to the rise of fully managed Kubernetes services over the past few years. The **Certified Kubernetes Administrator** (**CKA**) certification is designed to ensure that certification candidates have the skills and knowledge to help them establish their credibility and value in the job market, to support business growth.

This book will start with an introduction to the Kubernetes architecture and the core concept of Kubernetes, and then we will take a deep dive into main Kubernetes primitives with hands-on scenarios for installation and configuration, cluster management and workload scheduling, networking, and security. Furthermore, we'll discuss how to troubleshoot Kubernetes in our daily practice.

By the end of this book, you will be well versed in working with Kubernetes installation and configuration, and comfortable with the cluster management, storage, network, security-related configurations, and troubleshooting skills on vanilla Kubernetes.

If you want to learn more about Kubernetes, check out this playlist Kubernetes in 30 days - `https://www.youtube.com/watch?v=csPu6y6A7oY&list=PLyDI9q8xNovlhCqRhouXmSKQ-PP6_SsIQ`

Who this book is for

This book is targeted toward application developers, DevOps engineers, data engineers, and cloud architects who want to pass the CKA exam to certify their Kubernetes Administrator skills in the market. A basic knowledge of Kubernetes is recommended.

What this book covers

Chapter 1, *Kubernetes Overview*, introduces the Kubernetes architecture and its core concepts. It dives into common Kubernetes tools and gets hands-on with them, showing the big picture of different distributions and ecosystems of Kubernetes.

Chapter 2, *Installing and Configuring Kubernetes Clusters*, introduces the different configurations of Kubernetes and gets your hands dirty by setting up a Kubernetes cluster with a single worker node and multiple worker nodes using proper tooling.

Chapter 3, *Maintaining Kubernetes Clusters*, introduces the different approaches while maintaining Kubernetes clusters, and gets hands-on performing upgrades for Kubernetes clusters, backing up and restoring ETCD. This chapter covers 25% of the CKA exam content.

Chapter 4, Application Scheduling and Lifecycle Management, describes using Kubernetes deployments to deploy pods, scaling pods, performing rolling updates and rollbacks, resource management, and using ConfigMaps to configure pods. This chapter covers 15% of the CKA exam content.

Chapter 5, Demystifying Kubernetes Storage, discusses the core concept of Kubernetes storage for stateful workloads and shows how to configure applications with mounted storage and dynamically persistent storage. This chapter covers 10% of the CKA exam content.

Chapter 6, Securing Kubernetes, covers how Kubernetes authentication and authorization pattern works, then dives into Kubernetes **role-based access control** (**RBAC**). From there, we'll put managing the security of applications deployed on Kubernetes into perspective. This part is less than 5% of the CKA exam content.

Chapter 7, Demystifying Kubernetes Networking, describes using the Kubernetes networking model and core concepts, as well as how to configure Kubernetes networking on the cluster nodes and network policies, configuring Ingress controllers and Ingress resources, configuring and leveraging CoreDNS, as well as how to choose an appropriate container network interface plugin. This chapter covers 20% of the CKA exam content.

Chapter 8, Monitoring and Logging Kubernetes Clusters and Applications, describes how to monitor Kubernetes cluster components and applications, and how to get infrastructure-level, system-level, and application-level logs to serve as a source of log analytics or for further troubleshooting. Together with the next two chapters about troubleshooting cluster components and applications and troubleshooting Kubernetes security and networking, it covers 30% of the CKA exam content.

Chapter 9, Troubleshooting Cluster Components and Applications, describes the general troubleshooting approaches, and how to troubleshoot errors caused by cluster component failure and issues that occurred during the application deployments.

Chapter 10, Troubleshooting Security and Networking, follows on from *Chapter 9* and provides the general troubleshooting approaches to troubleshoot errors caused by RBAC restrictions or networking settings. In *Chapter 6*, we touched on how to enable Kubernetes RBAC and work with Kubernetes DNS. Be sure to go back and review those important concepts before diving into this chapter.

To get the most out of this book

This book is a comprehensive hands-on study guide focusing on providing hands-on skills with scenarios, and at the same time providing core knowledge to help readers warm up. The software and hardware covered in the book are as follows:

Software/hardware covered in the book	Operating system requirements
Minikube	Windows, macOS, or Linux
kubectl, kubeadm	Windows or Linux
Docker Desktop	Windows 10 or 11
WSL 2	Windows 10 or 11

Download the color images

We also provide a PDF file that has color images of the screenshots and diagrams used in this book. You can download it here: https://packt.link/AKr3r.

Conventions used

There are a number of text conventions used throughout this book.

Code in text: Indicates code words in text, database table names, folder names, filenames, file extensions, pathnames, dummy URLs, user input, and Twitter handles. Here is an example: "You can start by setting up an alias for kubectl using the alias k=kubectl command, and then use the k get command."

A block of code is set as follows:

```
apiVersion: v1
kind: Pod
metadata:
    name: melon-serviceaccount-pod
spec:
    serviceAccountName: melon-serviceaccount
    containers:
    - name: melonapp-svcaccount-container
      image: busybox
      command: ['sh', '-c','echo stay tuned!&& sleep 3600']
```

When we wish to draw your attention to a particular part of a code block, the relevant lines or items are set in bold:

```
spec:
    serviceAccountName: melon-serviceaccount
    containers:
```

Any command-line input or output is written as follows:

```
kubectl delete samelon-serviceaccount
```

Bold: Indicates a new term, an important word, or words that you see onscreen. For instance, words in menus or dialog boxes appear in **bold**. Here is an example: "This command will return the node that is now shown as **uncordoned**."

> **Tips or important notes**
> Appear like this.

Get in touch

Feedback from our readers is always welcome.

General feedback: If you have questions about any aspect of this book, email us at customercare@packtpub.com and mention the book title in the subject of your message.

Errata: Although we have taken every care to ensure the accuracy of our content, mistakes do happen. If you have found a mistake in this book, we would be grateful if you would report this to us. Please visit www.packtpub.com/support/errata and fill in the form.

Piracy: If you come across any illegal copies of our works in any form on the internet, we would be grateful if you would provide us with the location address or website name. Please contact us at copyright@packt.com with a link to the material.

If you are interested in becoming an author: If there is a topic that you have expertise in and you are interested in either writing or contributing to a book, please visit authors.packtpub.com.

Share Your Thoughts

Once you've read *Certified Kubernetes Administrator (CKA) Exam Guide* , we'd love to hear your thoughts! Scan the QR code below to go straight to the Amazon review page for this book and share your feedback.

https://packt.link/r/1803238267

Your review is important to us and the tech community and will help us make sure we're delivering excellent quality content.

Part 1: Cluster Architecture, Installation, and Configuration

This part looks at an overview of Kubernetes and explores its concepts and tooling. Furthermore, you will learn how to install and set up Kubernetes clusters. This part covers 25% of the CKA exam's content.

This part of the book comprises the following chapters:

- *Chapter 1, Kubernetes Overview*
- *Chapter 2, Installing and Configuring Kubernetes Clusters*
- *Chapter 3, Maintaining Kubernetes Clusters*

1

Kubernetes Overview

This chapter is an introduction to the Kubernetes architecture and Kubernetes core concepts. It dives into common Kubernetes tools and gets hands-on with them, showing the big picture of the different distributions and ecosystems in Kubernetes. In this chapter, we're going to cover the following main topics:

- CKA exam overview
- Cluster architecture and components
- Kubernetes core concepts
- Kubernetes in-market distribution and ecosystems

CKA exam overview

Certified Kubernetes Administrator (**CKA**) certification is a hands-on exam with a set of common Kubernetes working scenarios. You need to achieve it within a limited time frame. We highly recommend you work through this book within your environment and make sure that you understand and practice all the steps until you train your intuition and can perform all the tasks quickly without thinking twice. Time management is the key to success in this exam.

At the time of writing this book, the CKA exam is based on Kubernetes 1.22. Please check out the official example page to make sure you're up to date on any changes in the exam curriculum: `https://www.cncf.io/certification/cka/`. To learn more about the changes in Kubernetes, please check out the community release notes: `https://github.com/kubernetes/kubernetes/releases`.

The content of this book is well aligned with the CKA exam curriculum:

- *Part 1 – Chapters 1* to *3* cover *Kubernetes Cluster Architecture, Installation, and Configurations,* which makes up about 25% of the exam.

- *Part 2 – Chapter 4* covers *Workloads and Scheduling,* which makes up about 15% of the exam, *chapter 5* covers *Storage Services and Networking,* which makes up about 10% of the exam, *chapters 6* and *7* cover *Services and Networking,* which makes up about 20% of the exam.

- *Part 3 – Chapters 8* to *10* cover *Troubleshooting,* which makes up about 30% of the exam.

The goal of the exam curriculum is to help you prepare for the CKA exam and help you get a thorough understanding of each area, which will help you become skilled Kubernetes administrators later on in your career. While going through this book, please feel free to jump to the area that you need to know the most about if you're already familiar with some other topics.

Note that some Kubernetes security content before November 2020 has gradually moved to the **Certified Kubernetes Security Specialist (CKS)** exam. As a well-rounded Kubernetes administrator, it's essential to have a deep understanding of Kubernetes security. In fact, it is somewhat difficult to separate Kubernetes security as a different topic; however, knowledge of topics such as security context and **role-based access control (RBAC)** is still required for you to perform certain tasks to be successful in the exam. Therefore, this book will still cover some key security concepts to lay the groundwork if you want to pursue the CKS certification later on. To get to know more about different Kubernetes certifications, check out the FAQs from the Linux Foundation website by navigating to `https://docs.linuxfoundation.org/tc-docs/certification/faq-cka-ckad-cks`.

What to expect in your CKA exam

Prior to your exam, you have to make sure the computer you're going to use during the exam meets the system requirements defined by the exam provider. A webcam and microphone are mandatory to turn on during the exam. You're only allowed to use a single instance of a Chromium-based browser for the exam. You can find a list of Chromium-based browsers here: `https://en.wikipedia.org/wiki/Chromium_(web_browser)`.

Please make sure your hardware meets the minimum requirements by running the compatibility check tool, which you can find here: `https://www.examslocal.com/ScheduleExam/Home/CompatibilityCheck`. The detailed system requirements are defined here: `https://docs.linuxfoundation.org/tc-docs/certification/faq-cka-ckad-cks#what-are-the-system-requirements-to-take-the-exam`.

> **Important note**
> As this exam is an online remote-proctored exam, you can also check out what the exam is like here: `https://psi.wistia.com/medias/5kidxdd0ry`.

During your exam, you're allowed to check the official Kubernetes documentation including articles and documents under `https://kubernetes.io` and `https://github.com/kubernetes` on the same browser instance as the exam screen. The CKA exam consists of a set of around 20 scenario-based tasks to be achieved with a Linux-based shell and a set of predefined Kubernetes clusters. Those scenario-based tasks are described as a problem to be resolved with additional information. Candidates are bound to come up with the solutions based on the provided information and perform the solution promptly. A CKA exam session is about 2 hours, and after that, the exam will be marked as delivered. You can take the exam with multiple monitors if you wish to, although check out the exam policy beforehand to make sure you have met all the requirements from the organizer: `https://docs.linuxfoundation.org/tc-docs/certification/faq-cka-ckad-cks#how-is-the-exam-proctored`.

We highly recommend you walk through the sample scenarios provided by **killer.sh**, an official exam simulator, and bookmark the official documents that will be useful for you. Go to the **killer.sh** training website at `https://killer.sh/course/` to test out a simulated exam environment and test out the scenarios.

For more CKA exam instructions and tricks, please check out `https://docs.linuxfoundation.org/tc-docs/certification/tips-cka-and-ckad`.

You need a score of at least 66% to pass the exam, and the results will be emailed to you within 24 to 36 hours of finishing the exam. Accordingly, you will receive the certification in PDF form with a validity of 3 years, and a badge shortly after that. In case of any questions, you could email `certificationsupport@cncf.io` for further help.

CKA exam tips and tricks

Two key factors to help you succeed in the CKA exam or any other Kubernetes certifications are as follows:

- Excellent time management
- Practice, as we know that practice makes perfect

Before getting to the exam part, you have to be familiar with Kubernetes; don't dwell only on the certification when you're preparing for this exam. A deep understanding of the Kubernetes cluster architecture and ecosystem will help set a solid foundation on the way to learning any exam-related content.

Gaining some basic understanding of the Linux shell

Looking at the exam itself, a basic understanding of the Linux shell will assist you in achieving the goal quicker. The following commands will help you while you're going through the exercises in this book:

- `sudo` to avoid permission issues as much as possible, and `sudo su` to get root permission
- `curl`
- `| grep` in the command filtering result
- `vi/vim/nano` or other Linux text editor
- `cat`
- `cp/mv/mkdir/touch`
- `cp/scp`
- A good understanding of the `json` path is a plus, and using `jq` for JSON parsing would be a good complement to locating the information that you want to get out of the command.

As we're going through all the exam topics in this book, we'll cover most of these commands in the exercises. Make sure you understand and can confidently perform all the exercises independently with no rush.

Setting up a kubectl alias to save time

A lot of commands will be used repeatedly while you're working on various scenarios of the exam, so a friendly shortcut for `kubectl` is essential, as it will be used in nearly all of your commands:

```
alias k=kubectl
alias kg='kubectl get'
alias kgpo='kubectl get pod'
```

There's a `kubectl-aliases` repository on GitHub that you can refer to (`https://github.com/ahmetb/kubectl-aliases`). This was created by a contributor who showed some really good examples of `kubectl` aliases.

If you don't want to remember too much, you can try to understand the naming convention for shortcuts in Kubernetes. These would be things such as `svc` being short for services such that `kubectl get services` can become `kubectl get svc`, or `kubectl get nodes` can become `k get no`, for example. I have created a `melonkube playbook` repository, which covers all the shortcuts for Kubernetes objects (`https://github.com/cloudmelon/melonkube/blob/master/00%20-%20Shortcuts.md`).

You can refer to that to find what works best for you. However, please keep it simple as your mind may be get worked up during the actual exam for some reason. Practice and more practice will get you there sooner.

Setting kubectl autocomplete

You could set autocompletion in your shell; this will usually work in the Linux shell in your exam. You can achieve this with the following:

```
source <(kubectl completion bash) # setup autocomplete in bash
into the current shell, bash-completion package should be
installed first.
echo "source <(kubectl completion bash)" >> ~/.bashrc # add
autocomplete permanently to your bash shell.
```

Working in conjunction with the shortcut, you can do the following:

```
alias k=kubectl
complete -F __start_kubectl k
```

Although sometimes it may take more time to look for the right commands from bash autocompletion, I would say focusing on building a good understanding of the technology with practice will help you skill up faster.

Bookmarking unfamiliar yet important documentation in your browser

Get yourself familiar with Kubernetes official documentation to know where to find the information you need. The goal of CKA is *not* about memorizing but hands-on skills; knowing how to find the right path and resolving the challenge is the key. You could bookmark the documentation in the following domains:

- Kubernetes official documentation: https://kubernetes.io/docs/

- Kubernetes blog: https://kubernetes.io/blog/

- Kubernetes GitHub repository: https://github.com/kubernetes/

The first page that I usually recommend people to bookmark is the kubectl cheat sheet: https://kubernetes.io/docs/reference/kubectl/cheatsheet/. Another good bookmark is the official documentation search: https://kubernetes.io/search/?q=kubecon.

Be careful with the security context

The context is the most important indicator to let you know which Kubernetes cluster you're currently working on. We'll touch on the security context in more detail later in the book. I suggest you perform a context check before working on any new questions as you might get confused at times. Note that if you're not operating on the target Kubernetes cluster of that question, you will *not* get scored.

You can use the following command to check out the context:

```
kubectl config current-context
```

If you want to go to a specific Kubernetes cluster, you can use the following command:

```
kubectl config use-context my-current-cluster-name
```

You can also check out a list of Kubernetes clusters you've worked on with the following command in the actual exam:

```
kubectl config get-contexts
```

Managing your time wisely

Time management is the key to success in the CKA exam, and it is important to manage your time wisely by switching the task order. In general, all exam tasks are leveled from easy to difficult. When you reach the last few questions, you may find some tasks are quite time-consuming, but not the most difficult. You can skip to other questions that you're confident about and then come back to these later. That's why it's important to be aware of the Kubernetes cluster that you're currently working on.

Final thoughts

If you have walked through all the exercises in this book and want to gain a deeper understanding of Kubernetes, I recommend checking out another book that I co-authored back in 2020, called *The Kubernetes Workshop*, also published by Packt, which provides lots of Kubernetes exercises to help you skill up on the technology.

Cluster architecture and components

Kubernetes is a portable, highly extensible, open source orchestration that facilitates managing containerized workloads and services and orchestrates your containers to achieve the desired status across different worker nodes. It is worth mentioning that official documentation states that Kubernetes means *pilot* in Greek where its name originates from, which is appropriate for its function.

It supports a variety of workloads, such as stateless, stateful, and data-processing workloads. Theoretically, any application that can be containerized can be up and running on Kubernetes.

A Kubernetes cluster consists of a set of worker nodes; those worker machines run the actual workloads that are the containerized applications. A Kubernetes cluster can have from 1 up to 5,000 nodes (as of writing this chapter, we're on the Kubernetes 1.23 version).

We usually spin up one node for quick testing, whereas, in production environments, a cluster has multiple worker nodes for high availability and fault torrent.

Kubernetes adopts a master/worker architecture, which is a mechanism where one process acts as the master component to control one or more other components called slaves, or in our case, worker nodes. A general Kubernetes cluster architecture would look like the following:

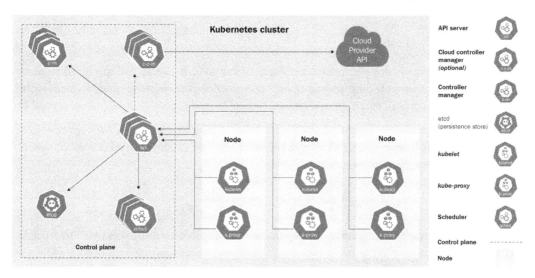

Figure 1.1 – Kubernetes cluster architecture

The Kubernetes master node, or the control plane, is in charge of responding to the cluster events, and it contains the following components:

- **API server**: This is the core of the Kubernetes control plane. The main implementation of the API server, also known as `kube-apiserver`, is to expose the Kubernetes REST API. You can see it as a communication manager between different Kubernetes components across the Kubernetes cluster.

- **etcd**: This is a distributed key-value store that stores information regarding the cluster information and all states of objects running on the Kubernetes cluster, such as Kubernetes cluster nodes, Pods, config maps, secrets, service accounts, roles, and bindings.

- **Kubernetes scheduler**: A Kubernetes scheduler is a part of the control plane. It is responsible for scheduling Pods to the nodes. `kube-scheduler` is the default scheduler for Kubernetes. You can imagine it as a postal officer who sends the Pod's information to each node and when it arrives at the target node, the `kubelet` agent on that node will provide the actual containerized workloads with the received specification.

- **Controllers**: Controllers are responsible for running Kubernetes toward the desired states. A set of built-in controllers runs inside `kube-controller-manager` in Kubernetes. Examples of those controllers are replication controllers, endpoint controllers, and namespace controllers.

Besides the control plane, every worker node in a Kubernetes cluster running the actual workloads has the following components:

- **kubelet**: A kubelet is an agent that runs on each worker node. It accepts pod specifications sent from the API server or locally (for static pod) and provisions the containerized workloads such as the Pod, StatefulSet, and ReplicaSet on the respective nodes.

- **Container runtime**: This is the software virtualization layer that helps run containers within the Pods on each node. Docker, CRI-O, and containerd are examples of common container runtimes working with Kubernetes.

- **kube-proxy**: This runs on each worker node and implements the network rules and traffic forwarding when a service object is deployed in the Kubernetes cluster.

Knowing about those components and how they work will help you understand the core Kubernetes core concepts.

Kubernetes core concepts

Before diving into the meat and potatoes of Kubernetes, we'll explain some key concepts in this section to help you start the journey with Kubernetes.

Containerized workloads

A containerized workload means applications running on Kubernetes. Going back to the raw definition of containerization, a container provides an isolated environment for your application, with higher density and better utilization of the underlying infrastructure compared to the applications deployed on the physical server or **virtual machines (VMs)**:

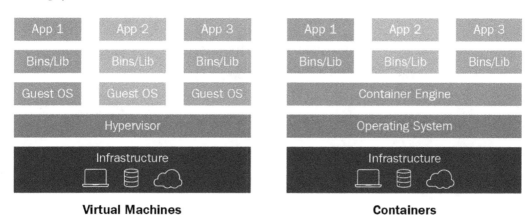

Figure 1.2 – Virtual machine versus containers

The preceding diagram shows the difference between VMs and containers. When compared to VMs, containers are more efficient and easier to manage.

Container images

A container isolates the application with all its dependencies, libraries, binaries, and configuration files. The package of the application, together with its dependencies, libraries, binaries, and configurations, is what we call a **container image**. Once a container image is built, the content of the image is immutable. All the code changes and dependencies updates will need to build a new image.

Container registry

To store the container image, we need a container registry. The **container registry** is located on your local machine, on-premises, or sometimes in the cloud. You need to authenticate to the container registry to access its content to ensure security. Most public registries, such as DockerHub and quay.io, allow a wide range of non-gated container image distributions across the board:

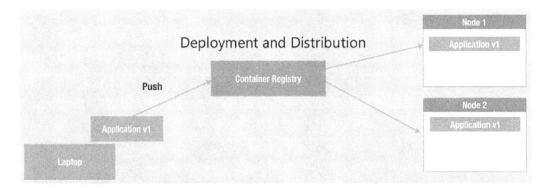

Figure 1.3 – Container images

The upside of this entire mechanism is that it allows the developers to focus on coding and configuring, which is the core value of their job, without worrying about the underlying infrastructure or managing dependencies and libraries to be installed on the host node, as shown in the preceding diagram.

Container runtimes

The container runtime is in charge of running containers, which is also known as the **container engine**. This is a software virtualization layer that runs containers on a host operating system. A container runtime such as Docker can pull container images from a container registry and manage the container life cycle using CLI commands, in this case, Docker CLI commands, as the following diagram describes:

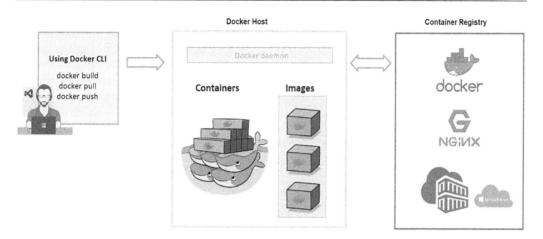

Figure 1.4 – Managing Docker containers

Besides Docker, Kubernetes supports multiple container runtimes, such as containerd and CRI-O. In the context of Kubernetes, the container runtime helps get containers up and running within the Pods on each worker node. We'll cover how to set up the container runtime in the next chapter as part of preparation work prior to provisioning a Kubernetes cluster.

> **Important note**
> Kubernetes runs the containerized workloads by provisioning Pods run on worker nodes. A node could be a physical or a virtual machine, on-premises, or in the cloud.

Kubernetes basic workflow

We saw earlier a typical workflow showing how Kubernetes works with Kubernetes components, and how they collaborate with each other, in the *Cluster architecture and components* section. When you're using `kubectl` commands, a YAML specification, or another way to invoke an API call, the API server creates a Pod definition and the scheduler identifies the available node to place the new Pod on. The scheduler does two things: *filtering* and *scoring*. The filtering step finds a set of available candidate nodes to place the Pod, and the scoring step ranks the most fitting Pod placement.

The API server then passes that information to the kubelet agent on the target worker node. The kubelet then creates the Pod on the node and instructs the container runtime engine to deploy the application image. Once it's done, the kubelet communicates the status back to the API server, which then updates the data in the `etcd` store, and the user will be notified that the Pod has been created.

This mechanism is repeated every time we perform a task and talk to the Kubernetes cluster, either by using `kubectl` commands, deploying a YAML definition file, or using other ways to trigger a REST API call through the API server.

The following diagram shows the process that we just described:

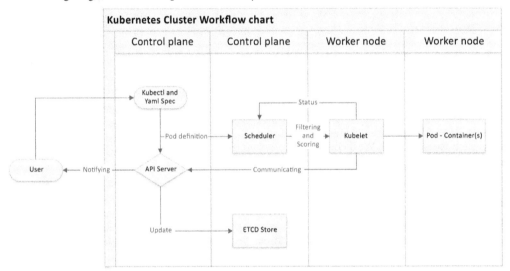

Figure 1.5 – Kubernetes cluster basic workflow

Knowing the basic Kubernetes workflow will help you understand how Kubernetes cluster components collaborate with each other and lay the foundation for learning about the Kubernetes plugin model and API objects.

Kubernetes plugin model

One of the most important reasons for Kubernetes to dominate the market and become the new normal of the cloud-native ecosystem is that it is flexible, highly configurable, and has a highly extensible architecture. Kubernetes is highly configurable and extensible on the following layers:

- **Container runtime**: The container runtime is the lowest software virtualization layer running containers. This layer supports a variety of runtimes in the market thanks to the **Container Runtime Interface** (**CRI**) plugin. The CRI contains a set of protocol buffers, specifications, a gRPC API, libraries, and tools. We'll cover how to cooperate with different runtimes when provisioning the Kubernetes cluster in *Chapter 2, Installing and Configuring Kubernetes Clusters*.

- **Networking**: The networking layer of Kubernetes is defined by kubenet or the **Container Network Interface** (**CNI**), which is responsible for configuring network interfaces for Linux containers, in our case, mostly Kubernetes Pods. The CNI is actually a **Cloud Native Computing Foundation** (**CNCF**) project that includes CNI specifications, plugins, and libraries. We'll cover more details about Kubernetes networking in *Chapter 7, Demystifying Kubernetes Networking*.

- **Storage**: The storage layer of Kubernetes was one of the most challenging parts at a time prior to **Container Storage Interface** (**CSI**) being introduced as a standard interface for exposing block and file storage systems. The storage volumes are managed by storage drivers tailored by storage vendors, this part previously being part of Kubernetes source code. The CSI compatible volume drivers are served for users to attach or mount the CSI volumes to the Pods running in the Kubernetes cluster. We'll cover storage management in Kubernetes in *Chapter 5, Demystifying Kubernetes Storage*.

They can be easily laid out as shown in the following diagram:

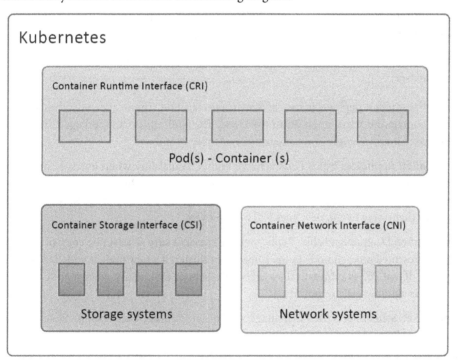

Figure 1.6 – Kubernetes plugin model

A good understanding of the Kubernetes plugin model will help you not only in your daily work as a Kubernetes administrator but also to lay the foundation to help you quickly learn about Kubernetes ecosystems and cloud-native community standards.

Kubernetes API primitives

All operations and communications between components and external user commands are REST API calls that the API server handles. Everything in Kubernetes is considered an API object.

In Kubernetes, when you run a `kubectl` command, the `kubectl` utility is in fact reaching kube-apiserver. `kube-apiserver` first authenticates and validates requests and then updates information in `etcd` and retrieves the requested information.

When it comes down to each worker node, the kubelet agent on each node takes Podspecs that are primarily provided by the API server, provisions the containerized workloads, and ensures (as described in those Podspecs) that the Pods are running and healthy. A Podspec is the body of the YAML definition file, which is translated to a JSON object that describes the specification for the workloads. Kubernetes form an API call going through the API server. And it is then taken into consideration by the control plane.

Kubernetes API primitives, also known as Kubernetes objects, are the fundamental building blocks of any containerized workload up and running in the Kubernetes cluster.

The following are the main Kubernetes objects we're going to use in our daily life while working with Kubernetes clusters:

- **Pods**: The smallest deployable unit in Kubernetes is a Pod. The worker node hosts the Pods, which contain the actual application workload. The applications are packaged and deployed in the containers. A single Pod contains one or more containers.

- **ReplicaSet**: ReplicaSet helps Pods achieve higher availability when users define a certain number of replicas at a time with a ReplicaSet. The role of the ReplicaSet is to make sure the cluster will always have an exact number of replicas up and running in the Kubernetes cluster. If any of them were to fail, new ones will be deployed.

- **DaemonSet**: DaemonSet is like ReplicaSet but it makes sure at least one copy of your Pod is evenly presented on each node in the Kubernetes cluster. If a new node is added to the cluster, a replica of that Pod is automatically assigned to that node. Similarly, when a node is removed, the Pod is automatically removed.

- **StatefulSet**: StatefulSet is used to manage stateful applications. Users can use StatefulSet when a storage volume is needed to provide persistence for the workload.

- **Job**: A job can be used to reliably execute a workload automatically. When it completes, typically, a job will create one or more Pods. After the job is finished, the containers will exit and the Pods will enter the `Completed` status. An example of using jobs is when we want to run a workload with a particular purpose and make sure it runs once and succeeds.

- **CronJob**: CronJobs are based on the capability of a job by adding value to allow users to execute jobs on a schedule. Users can use a `cron` expression to define a particular schedule per requirement.

- **Deployment**: A Deployment is a convenient way where you can define the desired state Deployment, such as deploying a ReplicaSet with a certain number of replicas, and it is easy to roll out and roll back to the previous versions.

We'll cover more details about how to work with those Kubernetes objects in *Chapter 4, Application Scheduling and Lifecycle Management*. Stay tuned!

Sharing a cluster with namespaces

Understanding the basic Kubernetes objects will give you a glimpse of how Kubernetes works on a workload level, and we'll cover more details and other related objects as we go. Those objects running on the Kubernetes cluster will work just fine when we're doing the development or test ourselves or a quick onboarding exercise, although we'll need to think about the separation of the workloads when it comes to the production environment for those enterprise-grade organizations. That's where the namespace comes in.

A namespace is a logical separation of all the namespaced objects deployed in a single Kubernetes cluster. Examples of namespaced objects are Deployments, Services, Secrets, and more. Some other Kubernetes objects are cluster-wide, such as StorageClasses, Nodes, and PersistentVolumes. The name of a resource has to be unique within a namespace, but it's labeled by a namespace name and an object name across all namespaces.

Namespaces are intended to separate cluster resources between multiple users, which creates the possibility of sharing a cluster for multiple projects within an organization. We call this model the **Kubernetes multi-tenant model**. The multi-tenant model is an effective way to help different projects and teams share the cluster and get the most use out of the same cluster. The multi-tenant model helps minimize resource wasting. It comes in handy in particular when working with Kubernetes in the cloud as there is always a reservation of resources by the cloud vendors. Despite all the upsides, the multi-tenant model is also bringing extra challenges to resource management and security aspects. We'll cover resource management in *Chapter 4, Application Scheduling and Lifecycle Management*.

For better physical isolation, we recommend that organizations use multiple Kubernetes clusters. It will bring a physical boundary for different projects and teams, although the resources reserved by the Kubernetes system are also replicated across clusters. Beyond that, working across different Kubernetes clusters is also challenging, as it involves setting up an effective mechanism by switching the security context, as well as dealing with the complexity of the networking aspects. We'll cover Kubernetes security in *Chapter 6*, *Securing Kubernetes*, and Kubernetes networking in *Chapter 7*, *Demystifying Kubernetes Networking*. The following is a diagram showing a Kubernetes multi-tenancy and multi-cluster comparison:

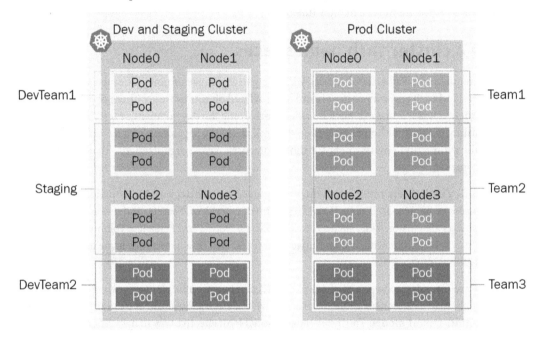

Figure 1.7 – Kubernetes multi-tenancy versus multi-cluster

Kubernetes in-market distribution and ecosystems

Kubernetes is supported by a fast-growing and vibrant open source community. There are more than 60 known Kubernetes platforms and distributions on the market. On the high level, there are managed Kubernetes and standard Kubernetes distributions from the upstream community. We're covering a high-level wrap-up for Kubernetes and its ecosystem in this section.

Upstream vanilla Kubernetes

Upstream vanilla Kubernetes is commonly used when the organization wants to manage the Kubernetes cluster and their own on-premises infrastructure or their cloud-based VM. The source code of Kubernetes distribution comes from the upstream Kubernetes community project. It's open for contribution, so feel free to join any **Special Interest Group** (**SIG**) groups; here's the full list of community groups: `https://github.com/kubernetes/community/blob/master/sig-list.md`.

If you have any ideas to share or want to learn from the community: `https://kubernetes.io/docs/contribute/generate-ref-docs/contribute-upstream/`.

Managed Kubernetes

Cloud vendor-managed Kubernetes distribution often falls into this category. Managed Kubernetes distribution is usually based on the vanilla Kubernetes cluster, and different vendors build their features on top of that and make it more adaptive to their infrastructure. A managed Kubernetes distribution usually has a control plane managed by the vendor, and users only need to manage the worker nodes and focus their energy on delivering value based on their core expertise.

Microsoft Azure provides **Azure Kubernetes Service** (**AKS**), **Amazon Web Service** (**AWS**) has **Elastic Kubernetes Service** (**EKS**), and **Google Cloud Platform** (**GCP**) is proud of its **Google Kubernetes Engine** (**GKE**).

Other popular Kubernetes distributions include VMware's Tanzu, RedHat OpenShift, Canonical's Charmed Kubernetes, and Kubernetes from Ranger Lab.

Kubernetes ecosystems

The Kubernetes ecosystem is not limited to provisioning and management tools; it has a wide variety of tools for security, networking, observability, and more. It covers all the important aspects of working with Kubernetes. The Kubernetes ecosystem is an important part of the cloud-native landscape. Thanks to Kubernetes being highly portable and platform-agnostic, we can literally take Kubernetes anywhere. It is easy to integrate with a security-sensitive disconnected scenario or integrated with the hybrid scenario as organizations are moving to the cloud. Those tools in the ecosystem are complementary to each other to boost Kubernetes' tremendous growth as a cloud-native technology and make a positive impact in the community and on the different sizes of businesses. Check out the cloud-native landscape at `https://landscape.cncf.io`.

Learning about Kubernetes and its ecosystem will help you better understand how to work with Kubernetes for your organization and how to help your organization get the best out of Kubernetes.

Summary

This chapter introduced you to some of the core concepts of Kubernetes, and we took a glimpse at the big picture of all the popular Kubernetes distributions on the market. An exciting journey is about to start!

In the next chapter, we'll dive into the details of the installation and configuration of a Kubernetes cluster. Stay tuned!

2

Installing and Configuring Kubernetes Clusters

This chapter introduces the different configurations of Kubernetes, which is the first step toward working with Kubernetes. We'll get our hands dirty by setting up a Kubernetes cluster with a single worker node and then multiple worker nodes. This chapter familiarizes you with Kubernetes installations, which is one of the key skills that will serve in your daily job as a Kubernetes administrator.

In this chapter, we're going to cover the following topics:

- Hands-on Kubernetes tooling

- Installing and configuring a Kubernetes cluster

- Using `minikube` to set up a single node Kubernetes cluster

- Using `kubeadm` to install a basic Kubernetes cluster

- Setting up a highly available cluster with `kubeadm`

Technical requirements

To get started, we need to make sure your local machine meets the technical requirements described as the following:

- A compatible Linux host – we recommend a Debian-based Linux distribution such as Ubuntu 18.04 or later.

- Make sure your host machine has at least 2 GB RAM, 2 CPU cores, and about 20 GB of free disk space.

Hands-on Kubernetes tooling

There are a handful of Kubernetes tools on the market – we'll start by covering some widely used Kubernetes tools to interact with the Kubernetes cluster. We'll dive into some key tools with hands-on labs later in this chapter.

Core tools

In this section, we are going to cover tools which are required to work with Kubernetes and containers.

kubectl

kubectl is a Kubernetes command-line tool used to talk to the Kubernetes cluster. It is hands down the most common and important utility that allows you to run commands against the Kubernetes cluster. There are a handful of kubectl commands available that will allow users to work with the Kubernetes cluster, such as deploying a containerized application, managing cluster resources, and monitoring and visualizing events and logs. We'll cover most of the common kubectl commands with examples as we go through the process.

To set up the kubectl utility, if you're on Red Hat-based distributions such as CentOS or Fedora, check out the official article for further information: https://kubernetes.io/docs/tasks/tools/install-kubectl-linux/#install-using-native-package-management. You can use the following commands:

```
cat <<EOF | sudo tee /etc/yum.repos.d/kubernetes.repo
[kubernetes]
name=Kubernetes
baseurl=https://packages.cloud.google.com/yum/repos/kubernetes-
el7-x86_64
enabled=1
gpgcheck=1
repo_gpgcheck=1
gpgkey=https://packages.cloud.google.com/yum/doc/yum-key.gpg
https://packages.cloud.google.com/yum/doc/rpm-package-key.gpg
EOF
sudo yum install -y kubectl
```

If you're on Debian-based distributions such as Ubuntu 18.04, you can follow the following instructions:

1. Firstly, you need to update the `apt` package index – then, you need to install the packages needed to use the Kubernetes `apt` repository by running the following commands sequentially:

    ```
    sudo apt-get update
    sudo apt-get install -y apt-transport-https
    ca-certificates curl
    ```

2. Download the Google Cloud public signing key and add the Kubernetes `apt` repository by using the following command:

    ```
    sudo curl -fsSLo /usr/share/keyrings/kubernetes-archive-
    keyring.gpg https://packages.cloud.google.com/apt/doc/
    apt-key.gpg
    echo "deb [signed-by=/usr/share/keyrings/kubernetes-
    archive-keyring.gpg] https://apt.kubernetes.io/
    kubernetes-xenial main" | sudo tee /etc/apt/sources.
    list.d/kubernetes.list
    ```

3. Now, you're ready to go. Make sure you update the `apt` package index with the new repository again and then install the `kubectl` utility using the `apt-get install` command:

    ```
    sudo apt-get update
    sudo apt-get install -y kubectl
    ```

4. You can verify whether `kubectl` has been successfully installed by running the following command upon the completion of the previous steps:

    ```
    kubectl version --client
    ```

    You'll see an output similar to the following if you have installed `kubectl` successfully:

```
cloudmelon@cloudmelonsrv:~$ kubectl version --client
Client Version: version.Info{Major:"1", Minor:"23", GitVersion:"v1.23.1", GitCommit:"86ec240af8cbd1b60bcc4c83c20da9b98085b92e", GitTreeState:"clean", Buil
dDate:"2021-12-16T11:41:01Z", GoVersion:"go1.17.5", Compiler:"gc", Platform:"linux/amd64"}
```

Figure 2.1 – A successful installation of kubectl

For instructions on installing `kubectl` in different environments, please refer to https://kubernetes.io/docs/tasks/tools/.

Container runtimes

Now, we are going to set up `containerd` as our container runtime by following these instructions:

1. Update the `apt` index, add Docker's official GPG key, and set up the `apt` repository by running the following instructions:

    ```
    sudo apt-get update
    sudo apt-get install \
        ca-certificates \
        curl \
        gnupg \
        lsb-release
    curl -fsSL https://download.docker.com/linux/ubuntu/gpg |
    sudo gpg --dearmor -o /usr/share/keyrings/docker-archive-
    keyring.gpg
    echo \
      "deb [arch=$(dpkg --print-architecture) signed-by=/
    usr/share/keyrings/docker-archive-keyring.gpg] https://
    download.docker.com/linux/ubuntu \
      $(lsb_release -cs) stable" | sudo tee /etc/apt/sources.
    list.d/docker.list > /dev/null
    ```

2. Install the Docker engine and `containerd.io`:

    ```
    sudo apt-get update
    sudo apt-get install docker-ce docker-ce-cli containerd.
    io
    ```

3. Validate that Docker has been installed successfully by using the following commands:

    ```
    sudo docker ps
    #optional - running your first docker container
    sudo docker run hello-world
    ```

    You'll see an output similar to the following:

Figure 2.2 – Docker is up and running

4. If you're about to configure `containerd` as the container runtime, you can use the following command and set the configuration to `default`:

    ```
    sudo mkdir -p /etc/containerd
    containerd config default | sudo tee /etc/containerd/
    config.toml
    ```

5. Restart `containerd` to make sure the changes take effect:

    ```
    sudo systemctl restart containerd
    ```

If you want to know more about how to set up CRI-O as a runtime, please check out the following link: `https://kubernetes.io/docs/setup/production-environment/container-runtimes/#cri-o`. It will show you how `containerd` serves as a container runtime in the context of Kubernetes.

Deployment tools

To bootstrap a Kubernetes cluster, we rely on the deployment tools. There are lots of useful tools on the market to help spin up a Kubernetes cluster, of which a lot of them are vendor-affinity. Here, we will cover what's requested in the CKA exam. That's the primary reason that we focus on upstream Kubernetes and these tools will help bootstrap a cluster on-premises. The following tools help you set up a Kubernetes cluster and we'll cover the detailed instructions while working with each of them in the next chapter:

*   **kubeadm**: `kubeadm` is the most important tool to help you crack the exam exercises. It helps install and set up the Kubernetes cluster with best practices. With `kubeadm`, you can provision a single node cluster and, more importantly, multi-node clusters. This is the first choice for most large organizations that want to manage their own Kubernetes cluster and use their own on-premises servers.

*   **minikube**: `minikube` is a popular local Kubernetes that can be provisioned on your local laptop or a **virtual machine** (**VM**). It's very lightweight, focusing on making it easy to learn and testing Kubernetes quickly.

*   **kind**: `kind` is similar to `minikube`. It focuses on provisioning local Kubernetes clusters and some simple CI scenarios and development. It runs local Kubernetes clusters using a Docker runtime – it can run as a single node Kubernetes cluster or a Kubernetes multi-node cluster. You can test lots of useful, simple scenarios with `kind`.

Other tools

Some of the other tools are not covered in the CKA exam – however, they will still come in handy in your daily work as a Kubernetes administrator.

Helm

Helm is a management tool for managing packages of pre-configured Kubernetes objects in the form of charts – we call these Helm charts.

To install `helm`, you can follow the following instructions for a Debian-based distribution such as Ubuntu 18.04:

1. Update the `apt` package index:

    ```
    curl https://baltocdn.com/helm/signing.asc | sudo apt-key
    add -
    sudo apt-get install apt-transport-https --yes
    ```

2. Install the packages to use the Helm `apt` repository with the following command:

    ```
    echo "deb https://baltocdn.com/helm/stable/debian/ all
    main" | sudo tee /etc/apt/sources.list.d/helm-stable-
    debian.list
    ```

3. Make sure you update the `apt` package index with the new repository again and then install Helm using the `apt-get install` command:

    ```
    sudo apt-get update
    sudo apt-get install helm
    ```

4. Use the following Helm command to validate its successful installation:

    ```
    helm version
    ```

    You'll see output similar to the following:

```
cloudmelon@cloudmelonsrv:~$ helm version
version.BuildInfo{Version:"v3.7.2", GitCommit:"663a896f4a815053445eec4153677ddc24a8a361", GitTreeState:"clean", GoVersion:"go1.16.18"}
```

Figure 2.3 – Successful installation of Helm

To know more ways to install Helm, check out the following link: `https://helm.sh/docs/intro/install/`.

Kompose

Most people who work with Docker will know about Docker Compose. Docker Compose is a tool used to define and run the multi-container applications containerized by Docker. It also uses a YAML file to define the application specifications. As more and more people are moving away from purely using Docker Swarm or Docker Desktop to take advantage of the enterprise-scale container orchestration system, Kompose comes in handy as a conversion tool for Docker Compose to contain orchestrators such as Kubernetes – the same structure works for Redhat OpenShift too.

You can install Kompose by running the following instructions on your Ubuntu 18.04:

1. Fetch the `kompose` binary:

    ```
    curl -L https://github.com/kubernetes/kompose/releases/
    download/v1.26.0/kompose-linux-amd64 -o kompose
    chmod +x kompose
    sudo mv ./kompose /usr/local/bin/kompose
    ```

2. Then, you can fetch a `docker compose` example file from the official website and test the `kompose convert` command as follows:

    ```
    wget https://raw.githubusercontent.com/kubernetes/
    kompose/master/examples/docker-compose-v3.yaml -O docker-
    compose.yaml
    kompose convert
    ```

    Your output will look similar to the following:

```
cloudmelon@cloudmelonsrv:~$ kompose convert
INFO Kubernetes file "frontend-tcp-service.yaml" created
INFO Kubernetes file "redis-master-service.yaml" created
INFO Kubernetes file "redis-slave-service.yaml" created
INFO Kubernetes file "frontend-deployment.yaml" created
INFO Kubernetes file "redis-master-deployment.yaml" created
INFO Kubernetes file "redis-slave-deployment.yaml" created
```

Figure 2.4 – A kompose convert command translating Docker
compose into Kubernetes-native YAML-defined files

3. Then, deploy those YAML files to your local Kubernetes cluster by using the following command:

```
kubectl apply -f .
```

Your output will look similar to the following:

```
cloudmelon@cloudmelonsrv:~$ kubectl get po -w
NAME                          READY   STATUS    RESTARTS   AGE
frontend-68b574b85d-5czxb     1/1     Running   0          17m
redis-master-d9788d6d9-sk448  1/1     Running   0          17m
redis-slave-64f6b5454f-tkvx5  1/1     Running   0          17m
```

Figure 2.5 – Kubernetes Pods up and running

The preceding screenshot shows the Redis Pods running in your Kubernetes cluster.

The dashboard

You can install a web-based **user interface** (**UI**) to your Kubernetes cluster. It not only displays the cluster status and shows what's going on with the Kubernetes cluster but also allows you to deploy containerized applications, troubleshoot, and manage the cluster and all related resources in the cluster.

The following is a sample dashboard:

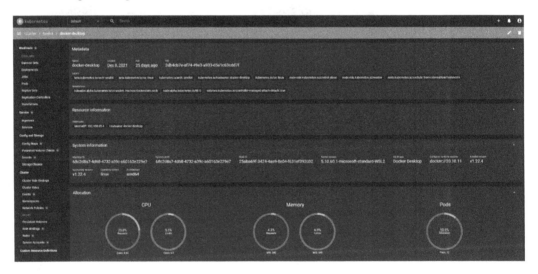

Figure 2.6 – The Kubernetes dashboard

The dashboard is sometimes handy for quick monitoring of the cluster states from the UI and user-friendly for collaborating with people who are not familiar with kubectl commands.

Installing and configuring a Kubernetes cluster

This section focuses on the installation of the Kubernetes cluster and the related configurations for it. With a good understanding gained from *Chapter 1*, where you learned about the Kubernetes cluster architecture and Kubernetes toolings, you will perform the Kubernetes cluster installation the hard way with `minikube` and `kubeadm`, and then update the cluster version.

Note that using `minikube` to spin up a single node cluster is not covered in the CKA exam but it comes quite handy when you'd like to test out Kubernetes in your local machine. The same goes for using `kubeadm` to install a Kubernetes multi-node cluster, as well as setting up a **highly available** (**HA**) Kubernetes cluster.

We expect you to learn both ways while putting more focus on the hands-on lab working with `kubeadm`. Starting with the next section, we'll walk you through the process of installing a new Kubernetes cluster and configuration.

Prerequisites for installing a Kubernetes cluster

To get started, we need to make sure your local machine meets the following technical requirements for both `minikube` and `kubeadm`:

- A compatible Linux host – we recommend a Debian-based Linux distribution such as Ubuntu 18.04 or later.

- Make sure your host machine has at least 2 GB RAM, 2 CPU cores, and about 20 GB of free disk space.

- Internet connectivity, as you will need to download dependencies throughout the process.

- A container runtime is needed prior to creating a Kubernetes cluster. During the cluster creation process, the Kubernetes cluster automatically detects an installed container runtime by scanning through the Unix domain sockets, if there are any, within your local machine. The **Unix domain socket** uses **Transmission Control Protocol** (**TCP**) as the underlying transport protocol. It is used for bidirectional data communication happening on the same operating system. We talked about how to install and configure container runtime in *Chapter 1* – please follow those instructions.

Before we get started, let's get the following checklist done.

Checking whether swap is disabled

For `kubeadm`, we have to disable `swap` in order to make `kubelet` work correctly, you can disable swap by doing the following:

```
sudo swapoff -a
```

Checking the container runtime

You can check the path to the Unix domain socket as instructed to verify your container runtime – this path is detectable by Kubernetes. Following the instructions to install Docker covered earlier in this chapter, you will find the Unix domain path under the /var/run/dockershim.sock path once you have installed the kubelet agent. To validate that Docker has been installed successfully, run the docker ps command:

```
sudo docker ps
```

The outcome of the following command is as follows:

```
[cloudmelon@cm-master-vm:~$ sudo docker ps
CONTAINER ID    IMAGE       COMMAND     CREATED     STATUS      PORTS       NAMES
```

Figure 2.7 – Checking the Docker runtime

If you have installed containerd as the container runtime, which we covered earlier in this chapter under the *Container runtimes* section, you will find the Unix domain path under the /run/containerd/containerd.sock path as the following:

Figure 2.8 – Checking the containerd runtime

kubeadm picks docker over containerd as the container runtime when both the docker and containerd runtimes are detected. At the time of writing, as announced at the beginning of Jan 2022, Kubernetes is removing dockershim in the upcoming v1.24 release. This is not surprising at all since it was first announced in Dec 2020 and Kubernetes' built-in dockershim component was deprecated in Kubernetes v1.20. In most cases, it won't affect the applications running in Kubernetes or the build process of the containerized applications if the following conditions are satisfied:

- There's no privileged root permission applied at the container level while it executes inside the pods using Docker commands and it restarts docker.service with systemctl
- Docker configuration files such as /etc/docker/daemon.json are modified

At this point, the official Kubernetes documentation has published this article to help users check whether dockershim deprecation will impact them. Check it out here for more ways to check the dependencies on Docker: https://kubernetes.io/docs/tasks/administer-cluster/ migrating-from-dockershim/check-if-dockershim-deprecation-affects- you/#find-docker-dependencies.

Checking whether the ports required by Kubernetes are opened

We also need to check if certain ports are open on your local machines prior to installing kubeadm. You can use the telnet command to do so:

```
telnet 127.0.0.1 6443
```

You can check the official documentation to make sure the ports and protocols used by Kubernetes are available by visiting this link: https://kubernetes.io/docs/reference/ports- and-protocols/.

Ensuring iptables sees bridged traffic

Make sure your Linux node's iptables is correctly configured to be able to watch the bridged traffic. You can set the net.bridge.bridge-nf-call-iptables parameter to a value of 1, just as we did here:

```
cat <<EOF | sudo tee /etc/modules-load.d/k8s.conf
br_netfilter
EOF

cat <<EOF | sudo tee /etc/sysctl.d/k8s.conf
net.bridge.bridge-nf-call-ip6tables = 1
net.bridge.bridge-nf-call-iptables = 1
EOF
sudo sysctl --system
```

You'll see an output similar to the following:

```
cloudmelon@cloudmelonplaysrv:~$ cat <<EOF | sudo tee /etc/modules-load.d/k8s.conf
br_netfilter
EOF

cat <<EOF | sudo tee /etc/sysctl.d/k8s.conf
net.bridge.bridge-nf-call-ip6tables = 1
net.bridge.bridge-nf-call-iptables = 1
EOF
sudo sysctl --system
br_netfilter
net.bridge.bridge-nf-call-ip6tables = 1
net.bridge.bridge-nf-call-iptables = 1
* Applying /etc/sysctl.d/10-console-messages.conf ...
kernel.printk = 4 4 1 7
* Applying /etc/sysctl.d/10-ipv6-privacy.conf ...
net.ipv6.conf.all.use_tempaddr = 2
net.ipv6.conf.default.use_tempaddr = 2
* Applying /etc/sysctl.d/10-kernel-hardening.conf ...
kernel.kptr_restrict = 1
* Applying /etc/sysctl.d/10-magic-sysrq.conf ...
kernel.sysrq = 176
* Applying /etc/sysctl.d/10-network-security.conf ...
net.ipv4.conf.default.rp_filter = 2
net.ipv4.conf.all.rp_filter = 2
* Applying /etc/sysctl.d/10-ptrace.conf ...
kernel.yama.ptrace_scope = 1
* Applying /etc/sysctl.d/10-zeropage.conf ...
vm.mmap_min_addr = 32768
* Applying /usr/lib/sysctl.d/50-default.conf ...
kernel.sysrq = 16
kernel.core_uses_pid = 1
net.ipv4.conf.default.rp_filter = 2
sysctl: setting key "net.ipv4.conf.all.rp_filter": Invalid argument
net.ipv4.conf.default.accept_source_route = 0
sysctl: setting key "net.ipv4.conf.all.accept_source_route": Invalid argument
net.ipv4.conf.default.promote_secondaries = 1
sysctl: setting key "net.ipv4.conf.all.promote_secondaries": Invalid argument
net.ipv4.ping_group_range = 0 2147483647
net.core.default_qdisc = fq_codel
fs.protected_hardlinks = 1
fs.protected_symlinks = 1
fs.protected_regular = 1
fs.protected_fifos = 1
* Applying /usr/lib/sysctl.d/50-pid-max.conf ...
kernel.pid_max = 4194304
* Applying /etc/sysctl.d/99-sysctl.conf ...
* Applying /etc/sysctl.d/k8s.conf ...
net.bridge.bridge-nf-call-ip6tables = 1
net.bridge.bridge-nf-call-iptables = 1
* Applying /etc/sysctl.conf ...
```

Figure 2.9 – iptables watching bridged traffic

The preceding screenshot shows the values in `iptables` have been updated.

Checking whether you have installed kubectl

`kubectl` is the command-line utility that you can use to talk to the Kubernetes cluster. Using the `kubectl version` command, you can verify whether `kubectl` has been successfully installed:

```
kubectl version --client
```

A successful installation will show an output similar to the following:

```
cloudmelon@cm-master-vm:~$ kubectl version --client
Client Version: version.Info{Major:"1", Minor:"23", GitVersion:"v1.23.2", GitCommit:"9d142434e3af351a628bffee3939e64c681afa4d", GitTreeState:"clean", BuildDate:"2022-01-19T17:35:46Z", GoVersion:"go1.17.5", Compiler:"gc", Platform:"linux/amd64"}
```

Figure 2.10 – Checking the kubectl version

Make sure you have completed the checklist in this section before moving on to the next section. These tools and requirements are essential and you may use them accordingly in the future.

Using minikube to set up a single node Kubernetes cluster

Creating a Kubernetes cluster using `minikube` is the easiest way to spin up a local Kubernetes cluster and it can be achieved in a matter of minutes. Here's what you need to do.

Installing minikube

Follow these steps to install `minikube`:

1. On your local or cloud-based Linux VM, use the `curl` command to retrieve the `minikube` binary, and then install it under `/usr/local/bin/minikube` as follows:

    ```
    curl -LO https://storage.googleapis.com/minikube/
    releases/latest/minikube-linux-amd64
    sudo install minikube-linux-amd64 /usr/local/bin/minikube
    ```

2. You can go to `/usr/local/bin/minikube` to check whether you have successfully installed the `minikube` binary before moving to the next steps or you can also check by typing the following command into the terminal:

    ```
    minikube --help
    ```

Using minikube to provision a single node Kubernetes cluster

Follow these steps to use minikube to provision a single node Kubernetes cluster:

1. When using `minikube` to provision a single node Kubernetes cluster, you can simply use the `minikube start` command:

    ```
    minikube start
    ```

2. You can also set up the CPU cores and memory to start your `minikube` cluster by adding a `--memory` and `--cpus` flag as follows:

    ```
    minikube start --memory 8192 --cpus 4
    ```

After the command is executed, it kicks off the `minikube` cluster provisioning process. You'll see an output similar to the following:

```
minikube v1.24.0 on Ubuntu 18.04
Using the docker driver based on user configuration
Starting control plane node minikube in cluster minikube
Pulling base image ...
Downloading Kubernetes v1.22.3 preload ...
    > preloaded-images-k8s-v13-v1...: 501.73 MiB / 501.73 MiB  100.00% 72.40 Mi
    > gcr.io/k8s-minikube/kicbase: 355.77 MiB / 355.78 MiB  100.00% 32.75 MiB p
Creating docker container (CPUs=2, Memory=2200MB) ...
Preparing Kubernetes v1.22.3 on Docker 20.10.8 ...
    ▪ Generating certificates and keys ...
    ▪ Booting up control plane ...
    ▪ Configuring RBAC rules ...
Verifying Kubernetes components...
    ▪ Using image gcr.io/k8s-minikube/storage-provisioner:v5
Enabled addons: storage-provisioner, default-storageclass
Done! kubectl is now configured to use "minikube" cluster and "default" namespace by default
```

Figure 2.11 – Spinning up a minikube cluster

By the end, you will see a message telling you we're ready to use the `minikube` Kubernetes cluster (as concluded in the preceding screenshot).

Verifying the minikube cluster installation

Your `minikube` cluster contains one node that serves as both the control plane and worker node. That means that once you have it set up, you can start to schedule workloads in your local Kubernetes cluster. You can use the following command to see whether the node is ready to use:

```
kubectl get node
```

You can also use the shortcut of this command:

```
alias k=kubectl
k get no
```

The output will show you the following:

- The status of the node and whether it's ready to use
- The role of that node
- The Kubernetes version
- The age of that node since it's been deployed

Here is the output:

```
cloudmelon@cloudmelonsrv:~$ kubectl get no
NAME        STATUS    ROLES                   AGE    VERSION
minikube    Ready     control-plane,master    38s    v1.22.3
```

Figure 2.12 – Checking the Docker runtime

Configuring the minikube cluster

If you'd like to configure the `minikube` cluster without reprovisioning a new one, you need to stop the `minikube` cluster using the `minikube stop` command.

The `minikube config set` command will help you apply the settings such as CPU and memory that you'll allocate to the `minikube` cluster. After configuring the `minikube` cluster, you need to start the `minikube` cluster and from there, you'll be working on the cluster with the new configurations.

Here's the process to configure `minikube` using more memory and CPUs:

```
minikube stop
minikube config set memory 8192
minikube config set cpus 4
minikube start
```

After that, you can continue to play with the `minikube` cluster. In case you have any questions about how the commands work, use the `minikube config - - help` command to get help.

Deleting a minikube cluster

The following command deletes all local Kubernetes clusters and all profiles:

```
minikube delete --all
```

What you learned from this section can be used repeatedly every time you need a local Kubernetes cluster. You can replicate what you have learned from this section for quick testing of the latest Kubernetes release for most of the new features featured in the release note: `https://github.com/kubernetes/kubernetes/releases`.

However, most enterprise-grade environments will not be satisfied with a single node cluster. They are mostly multi-node setups. In the next section, we will dive into creating a Kubernetes multi-node cluster with `kubeadm`.

Using kubeadm to install a basic Kubernetes cluster

In this section, we will create a multi-node Kubernetes cluster using kubeadm. The following are the steps we need to achieve the goal:

1. Install kubeadm.

2. Bootstrap a master node where your control plane will be located

3. Install the network plugins (we will get to the detailed supported plugins later in this chapter and use Calico as an example in that section).

4. Bootstrap the worker nodes.

5. Join the worker nodes to the control plane.

Before getting started, you need to make sure your master node meets all the technical requirements listed in this chapter.

We'll deploy a basic Kubernetes cluster by going through the steps described in this section, as shown in *Figure 2.7*:

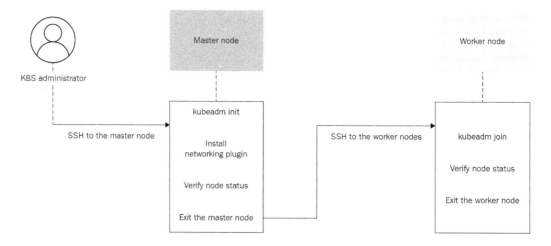

Figure 2.13 – The workflow of using kubeadm to spin up a basic Kubernetes cluster

The Kubernetes cluster will be similar to the architecture featured in *Figure 2.14*:

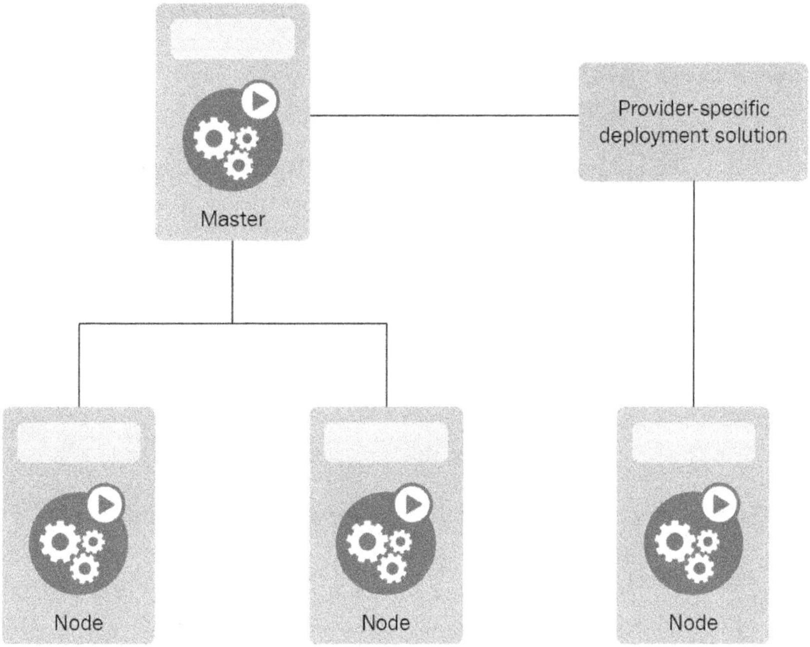

Figure 2.14 – A standard multi-node Kubernetes cluster

From now on, you can follow these instructions to create a multi-node Kubernetes cluster. To create a Kubernetes cluster using kubeadm, its default settings conform to best practices of setting up a standard Kubernetes cluster. This set of best practices is encapsulated as Kubernetes Conformance tests. Check out the details about the Kubernetes Conformance Program here: https://kubernetes. io/blog/2017/10/software-conformance-certification/.

Installing kubeadm

We introduced setting up docker or containerd as the container runtime – we can then install kubeadm by following these instructions:

1. Update the apt package index, add the Google Cloud public signing key, and set up the Kubernetes apt repository by running the following instructions:

    ```
    sudo apt-get update
    sudo apt-get install -y apt-transport-https
    ca-certificates curl
    sudo curl -fsSLo /usr/share/keyrings/kubernetes-archive-
    keyring.gpg https://packages.cloud.google.com/apt/doc/
    ```

```
apt-key.gpg
echo "deb [signed-by=/usr/share/keyrings/kubernetes-
archive-keyring.gpg] https://apt.kubernetes.io/
kubernetes-xenial main" | sudo tee /etc/apt/sources.
list.d/kubernetes.list
```

2. Start by updating the apt package index and then install kubelet and kubeadm:

```
sudo apt-get update
sudo apt-get install -y kubelet kubeadm
```

3. Here, if you haven't installed kubectl yet, you can also install kubelet, kubeadm, and kubectl in one go:

```
sudo apt-get update
sudo apt-get install -y kubelet kubeadm kubectl
```

4. Use the following command to pin the version of the utilities you're installing:

```
sudo apt-mark hold kubelet kubeadm kubectl
```

The output shows those packages are set on hold as shown in *Figure 2.9*:

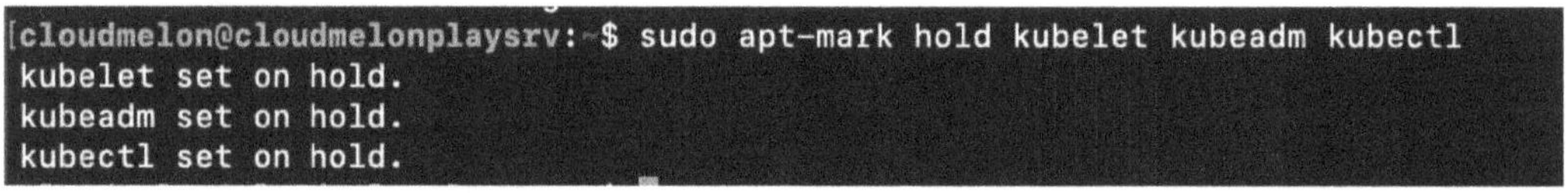

Figure 2.15 – Checking the containerd runtime

5. From here, you can check whether kubeadm has been successfully installed by typing kubeadm into the command shell. Here's the output of the command:

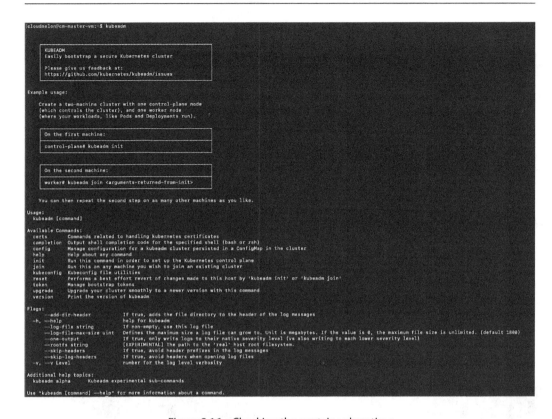

Figure 2.16 – Checking the containerd runtime

6. To verify that kubelet is present on the master node, you can use the which kubelet command, which returns the location of the kubelet agent:

```
[cloudmelon@cm-master-vm:~$ which kubelet
/usr/bin/kubelet
```

Figure 2.17 – Checking kubelet's presence

As you have successfully installed kubeadm and kubelet, you can now start initiating a control plane.

Here, we will show an optional operation where you can use images pull to pre-pull the images that are required to set up the Kubernetes cluster:

```
sudo kubeadm config images pull
```

The output should be similar to the following screenshot:

```
[cloudmelon@cm-master-vm:~$ sudo kubeadm config images pull
[config/images] Pulled k8s.gcr.io/kube-apiserver:v1.23.2
[config/images] Pulled k8s.gcr.io/kube-controller-manager:v1.23.2
[config/images] Pulled k8s.gcr.io/kube-scheduler:v1.23.2
[config/images] Pulled k8s.gcr.io/kube-proxy:v1.23.2
[config/images] Pulled k8s.gcr.io/pause:3.6
[config/images] Pulled k8s.gcr.io/etcd:3.5.1-0
[config/images] Pulled k8s.gcr.io/coredns/coredns:v1.8.6
```

Figure 2.18 – Pre-pulling the images

Note that the preceding operation is optional – you're free to skip it and go straight to the next section.

Bootstrapping a master node

You can use the kubeadm init command to initiate the control plane as a regular user and gain sudo privileges from your master node machine by using the following command:

```
sudo kubeadm init --pod-network-cidr=192.168.0.0/16
```

You will see an output similar to the following:

```
Your Kubernetes control-plane has initialized successfully!

To start using your cluster, you need to run the following as a regular user:

  mkdir -p $HOME/.kube
  sudo cp -i /etc/kubernetes/admin.conf $HOME/.kube/config
  sudo chown $(id -u):$(id -g) $HOME/.kube/config

Alternatively, if you are the root user, you can run:

  export KUBECONFIG=/etc/kubernetes/admin.conf

You should now deploy a pod network to the cluster.
Run "kubectl apply -f [podnetwork].yaml" with one of the options listed at:
  https://kubernetes.io/docs/concepts/cluster-administration/addons/

Then you can join any number of worker nodes by running the following on each as root:

kubeadm join 172.16.16.129:6443 --token k626hm.oqwyac35h43x80mg \
    --discovery-token-ca-cert-hash sha256:889983a6b87643e598b88533dbe3a68643a623b9a0ed9380561c6a7dbb93b3f0
```

Figure 2.19 – The control plane initiated successfully

After your Kubernetes control-plane is initialized successfully, you can execute the following commands to configure kubectl:

```
mkdir -p $HOME/.kube
sudo cp -i /etc/kubernetes/admin.conf $HOME/.kube/config
sudo chown $(id -u):$(id -g) $HOME/.kube/config
```

If you're a root user, you can use the following:

```
export KUBECONFIG=/etc/kubernetes/admin.conf
```

Then, the next step is to deploy a pod network to the Kubernetes cluster.

Installing the networking plugins

In order for the pods to talk to each other, you can deploy the networking by enabling **Container Network Interface** (**CNI**) plugin. The CNI plugins conform to the CNI specification, and as per the official Kubernetes documentation, Kubernetes follows the v0.4.0 release of the CNI specification.

There's a wide range of networking plugins working with Kubernetes – we will dive into Kubernetes networking in *Chapter 7, Demystifying Kubernetes Networking*. Here are some add-ons options:

- Calico
- Flannel
- Weave Net

For all the possible options acknowledged by the Kubernetes community, please check out the official documentation: https://kubernetes.io/docs/concepts/cluster-administration/addons/. You can check out the links from this page to get the installation instructions for the respective options.

Here, we're going to use the Calico plugin as the overlay network for our Kubernetes cluster. It is a Kubernetes CNI networking provider and it allows you to write up the network policies, which means that it supports a set of networking options to suit your different requirements. Here's how we'll approach it:

1. Deploy the Tigera Calico **Custom Resource Definitions** (**CRDs**) and operator by using the kubectl create -f command:

    ```
    kubectl create -f https://docs.projectcalico.org/
    manifests/tigera-operator.yaml
    kubectl create -f https://docs.projectcalico.org/
    manifests/custom-resources.yaml
    ```

2. You can use the watch command to monitor the pod status in the process:

    ```
    watch kubectl get pods -n calico-system
    ```

    Alternatively, use the following alternative command:

    ```
    kubectl get pods -n calico-system -w
    ```

Now, you can see the pods have a `Running` status:

```
Every 2.0s: kubectl get pods -n calico-system

NAME                                          READY   STATUS    RESTARTS   AGE
calico-kube-controllers-7dddfdd6c9-tpxv2      1/1     Running   0          51s
calico-node-f5cnd                             1/1     Running   0          51s
calico-typha-84bf84b9b7-tmk5x                 1/1     Running   0          51s
```

Figure 2.20 – The control plane initiated successfully

3. For the Kubernetes cluster created by `kubeadm`, there's a taint by default for master nodes. Therefore, we need to remove taints so that the master node is available to schedule pods. To remove the taint, you can use the following command:

    `kubectl taint nodes --all node-role.kubernetes.io/master-`

    The following screenshot shows that the taint on the master node has been successfully removed:

```
[cloudmelon@cloudmelonplaysrv:~$ kubectl taint nodes --all node-role.kubernetes.io/master-
node/cloudmelonplaysrv untainted
```

Figure 2.21 – Removing the taint on the master node successfully

4. You can use the following command to check out the current nodes that are available:

    `kubectl get no`

5. To get more information from the node, you can use the following command:

    `kubectl get no -o wide`

    The following screenshot shows the sample output:

```
cloudmelon@cloudmelonplaysrv:~$ kubectl get nodes -o wide
NAME                STATUS   ROLES                 AGE   VERSION   INTERNAL-IP     EXTERNAL-IP   OS-IMAGE      KERNEL-VERSION    CONTAINER-RUNTIME
cloudmelonplaysrv   Ready    control-plane,master  20h   v1.23.2   172.16.16.129   <none>        Ubuntu 21.10  5.13.0-23-generic docker://20.10.12
```

Figure 2.22 – The Kubernetes node status

From the preceding command output, you can see the Kubernetes node is operational after enabling the CNI networking and it has been assigned an internal IP address.

Bootstrapping the worker nodes

To add more worker nodes to the Kubernetes cluster, we will SSH to the client machine, and make sure the worker nodes meet the same technical requirements as the master node. Check out the *Prerequisites for installing a Kubernetes cluster* section of this chapter and refer to the information on kubeadm for more details. Make sure you have installed the container runtime and kubeadm, although kubectl is optional for worker nodes since we usually use the master node for management.

Joining the worker nodes to the control plane

We can go ahead with installing kubeadm for the master node after making sure that your worker nodes and local environment meet the technical requirements that we set, as we mentioned earlier in this section. As introduced in *Chapter 1, Kubernetes Overview*, the worker nodes are where your containerized workloads are up and running.

You can use the following command to join the worker nodes to the Kubernetes cluster. This command can be used repeatedly each time you have to join new worker nodes:

```
sudo kubeadm join --token <token> <control-plane-
host>:<control-plane-port> --discovery-token-ca-cert-hash
sha256:<hash>
```

You can actually go back and copy the output of the master node control plane, which would look similar to the following sample command:

```
sudo kubeadm join 172.16.16.129:6443 --token k626hm.
oqwyac35h43x80mg   --discovery-token-ca-cert-hash sha256:889983
a6b87643e598b88533dbe3a68643a623b9a0ed9380561c6a7dbb93b3f0
```

You can use the preceding command to join the worker node to the control plane and set up your Kubernetes cluster with multiple worker nodes.

Setting up a highly available cluster with kubeadm

In *Chapter 1, Kubernetes Overview*, we introduced the cluster architecture, which gives us two options: setting up a single node Kubernetes cluster for dev/test quick testing or setting up a multi-node Kubernetes cluster for more professional use, or even use in production. A standard configuration would be one master with multiple worker nodes. As we stated in the previous chapter, the Kubernetes master node is where the control plane resides. In the event of a master node going down, either the containerized workloads up and running in the worker nodes will still keep running until the worker node is off the grid for some reason or there are no available master nodes, meaning no new workloads will be scheduled to the worker node.

There are two options available to build a HA Kubernetes cluster:

- **Building multiple master nodes**: This is the option where the control plane nodes and etcd members co-exist in the same master nodes. *Figure 2.16* shows the stacked etcd topology:

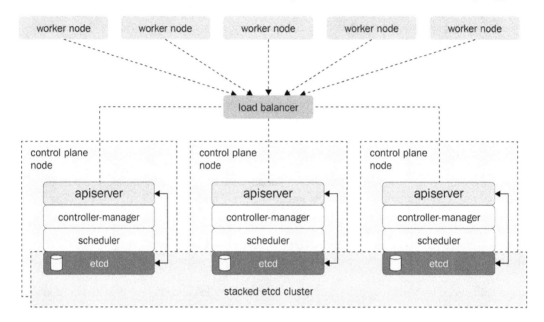

Figure 2.23 – A stacked etcd topology for a HA kubeadm cluster

This topology makes the cluster more resilient compared to the basic Kubernetes cluster architecture that we built in this chapter, thanks to the redundancy of the master node. In case one master node goes down, it's easy to switch to another available master node to ensure the health of the entire Kubernetes cluster.

However, in some cases where we need to manage the cluster and replicate the cluster information, the external etcd typology comes in.

- **Building an external etcd cluster**: Compared to the previous option, the key idea of this option is to decouple the etcd store to a separate infrastructure since the etcd, as we mentioned in *Chapter 1*, is where Kubernetes stores the cluster and the state information about the Kubernetes objects. The kubeadm HA topology architecture for an external etcd cluster is shown in *Figure 2.24*:

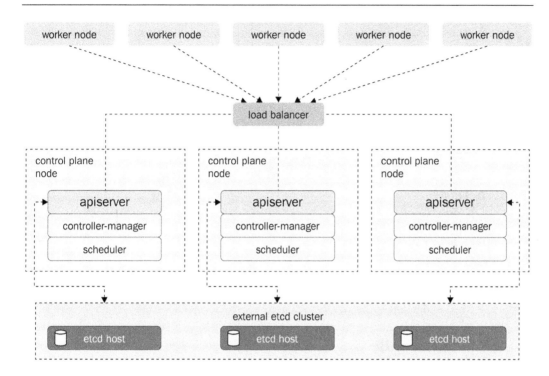

Figure 2.24 – The topology for an external etcd HA kubeadm cluster

As shown in *Figure 2.24*, the external etcd is a cluster and it communicates with the API server of each control plane. In the event of the control plane node going down, we won't lose all the information stored in the etcd store. It also makes the control plane more decoupled and manageable, as we only need to add more control plane nodes. A loss of the control plane node won't be as impactful as it would with the stacked etcd topology.

Summary

This chapter covers the very first job for most Kubernetes administrators who are setting up a Kubernetes cluster with a single worker node or with multiple worker nodes. The various tools introduced in this chapter will help your daily routine at work beyond the exam. Nevertheless, this is also one of the most time-consuming tasks in the CKA exam. Practice, practice, and more practice will help you get the hang of it. Knowing the HA topology for a Kubernetes cluster will also help you address the requirements of the organization that you'll be working for as a Kubernetes administrator. As you master the setup process for a basic Kubernetes cluster, it will become easier to apply your skills to different typologies.

In the next chapter, we'll talk about Kubernetes cluster maintenance, including some important topics such as upgrades to Kubernetes components, which is quite an essential task in the daily work of a Kubernetes administrator. Touching on external etcd typology in this chapter is just a start, as we'll dive into more interesting work with etcd in the next chapter. Happy learning!

Mock CKA scenario-based practice test

You have two VMs, *master-0* and *worker-0*. Please complete the following mock scenarios.

Scenario 1:

Install the latest version of `kubeadm`, then create a basic `kubeadm` cluster on the `master-0` node, and get the node information.

Scenario 2:

SSH to `worker-0` and join it to the `master-0` node.

Scenario 3 (optional):

Set up a local `minikube` cluster and schedule your first workload, called `hello Packt`

You can find all the scenario resolutions in *Appendix - Mock CKA scenario-based practice test resolutions* of this book.

FAQs

- *Where should I start to test the Kubernetes cluster?*

   You can start on your local laptop or desktop on Windows, Linux, or Mac OS, and we recommend using VMware player or Hyper-V to spin up multiple VMs so you can test out a multinode scenario. Using Multipass from Canonical is also great for creating Ubuntu VMs and it supports Linux, Mac, and Windows. Check it out here: `https://multipass.run/`.

   Another option is to get a cloud subscription such as Microsoft Azure, AWS, or GCP, using which you can provision a VM with a click-through experience.

- *Where can I find the latest Kubernetes release to test out?*

   The Kubernetes GitHub repository is where you can find all the releases as well as changelogs, and you can get the latest release and build it by yourself: `https://github.com/kubernetes/kubernetes`.

   We can also use `kubeadm` or `minikube` to get Kubernetes, as they are aligned with the Kubernetes source code delivery cycle and are up to date.

3

Maintaining Kubernetes Clusters

Kubernetes has been the most vibrant platform in the community over the past few years and it has maintained a good release cadence, which makes Kubernetes maintenance important in order to enable organizations that work with Kubernetes to take advantage of its latest features. This chapter introduces different approaches for maintaining a Kubernetes cluster while providing practical lessons on performing upgrades for Kubernetes clusters, etcd backup, and etcd restore. It covers 25% of the CKA exam content.

In this chapter, we're going to cover the following main topics:

- Demystifying Kubernetes cluster maintenance
- Performing a version upgrade on a Kubernetes cluster using kubeadm
- Working with etcd
- Backing up and restoring etcd

Demystifying Kubernetes cluster maintenance

Before April 2021, Kubernetes had maintained quite a steady and sound cadence of quarterly releases throughout the year. Despite the strong growth and incredible popularity in the community, this was reduced to three releases per year. Fewer releases still mean that a regular maintenance window should be scheduled within the organization for the upgrade of security patches, and to take full advantage of enhancements and new features.

A general maintenance window contains the upgraded Kubernetes cluster version. We can easily break the task to be performed into two parts:

- Upgrading the master node, which contains the control plane
- Upgrading the worker node

Upgrading the master node is simple if you have only one master node. However, most enterprise-grade customers may have a couple of master nodes for better resilience. We should be aware that working with an organization as a Kubernetes administrator is sometimes more challenging than working with other ones. A general reference of master node management would be the two typologies of high availability mentioned in *Chapter 2, Installing and Configuring Kubernetes Clusters*. The conventional way to upgrade the master node is to upgrade one at a time regardless of the number of master nodes in your current cluster. You will need to upgrade the kubeadm and kubectl versions.

When it comes to worker node upgrade, as we mentioned in the previous chapter, the worker node is where your workloads are actually running; therefore, you need to upgrade both the kubeadm and kubelet versions. Keep in mind that you need to upgrade one at a time when it comes to multiple worker nodes available in the current Kubernetes cluster.

If you have a separate etcd cluster set up, you will need to upgrade the etcd store version, which is not covered in the CKA exam. In general, you need to check out the official documentation to know more about Kubernetes components and version compatibility here: `https://kubernetes.io/releases/version-skew-policy/`.

Another general task for Kubernetes cluster maintenance is backup and restore with the etcd store. The etcd stores cluster data that includes cluster state information such as pod state data, node state data, and the configurations critical for Kubernetes. In most cases, as a Kubernetes administrator, you will need to perform the following two key tasks:

- Back up the etcd store regularly
- Restore the etcd from cluster failure

The following section will firstly walk you through the general process of upgrading the Kubernetes cluster with kubeadm. This is one of the most time-consuming questions in the actual CKA exam. Make sure you practice it a few times until you master the general upgrade process as well as how to perform upgrade tasks by following the official Kubernetes documentation. Note that update policies vary for managed Kubernetes distributions by cloud vendors. Please check their respective official documentation.

Furthermore, we'll take a look at how to back up and restore an etcd cluster.

Upgrading a Kubernetes cluster using kubeadm

Kubernetes versions follow semantic versioning, and are expressed in three parts:

1. Major version

2. Minor version

3. Patch version

For example, Kubernetes version 1.23.3 means that it is Kubernetes 1.23 minor version with patch number 3. Similarly, 1.22 and 1.21 are both minor versions like 1.23.

At the time of writing this book, Kubernetes 1.19+ has one year of path support. That means for Kubernetes 1.23, released in January 2022, the end of support will be February 2023. For Kubernetes 1.8 and older, the support patch was shortened to roughly 9 months instead. **Special interest group (SIG)** releases manage the Kubernetes release cycle, and the best way to keep track of the release schedule is to follow them at `https://github.com/kubernetes/sig-release/tree/master/releases` and read the change log at `https://github.com/kubernetes/kubernetes/tree/master/CHANGELOG` to keep up to date.

If you would like to upgrade a cluster to a targeted version, check out supported versions at `https://kubernetes.io/releases/version-skew-policy/#supported-versions`.

Upgrading the master node

Before you upgrade the master node, make sure you know the purpose of your upgrade and have backed up any important components. It is recommended that you start with checking out the current version and then determining which version to upgrade to. Once decided, we'll perform the following actions to upgrade the master node as depicted in *Figure 2.1*, including upgrading with kubeadm and interacting with Kubernetes nodes:

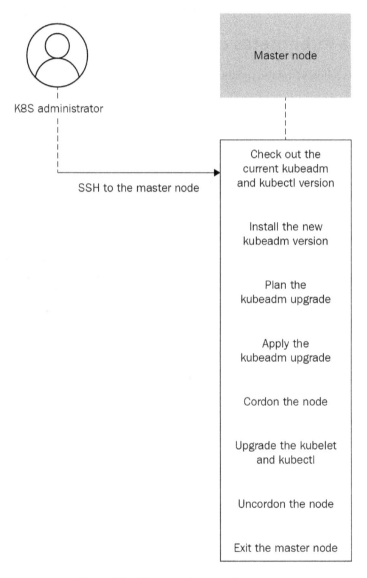

Figure 3.1 – Master node upgrade process

Let's start by checking out the current version with the following commands once we're in the master node:

```
kubeadm version
kubectl version
```

From the output, we know that we are currently on Kubernetes 2.23.2:

```
Client Version: version.Info{Major:"1",
Minor:"23", GitVersion:"v1.23.2",
GitCommit:"e6c093d87ea4cbb530a7b2ae91e54c0842d8308a",
GitTreeState:"clean", BuildDate:"2022-02-16T12:38:05Z",
GoVersion:"go1.17.7", Compiler:"gc", Platform:"linux/amd64"}
```

Let's check out the latest versions available with the following commands:

```
apt update
apt-cache madison kubeadm
```

Now we know the latest versions available:

```
[cloudmelon@cloudmelonplaysrv:~$ apt-cache madison kubeadm
   kubeadm |   1.23.3-00 | https://apt.kubernetes.io kubernetes-xenial/main arm64 Packages
   kubeadm |   1.23.2-00 | https://apt.kubernetes.io kubernetes-xenial/main arm64 Packages
   kubeadm |   1.23.1-00 | https://apt.kubernetes.io kubernetes-xenial/main arm64 Packages
   kubeadm |   1.23.0-00 | https://apt.kubernetes.io kubernetes-xenial/main arm64 Packages
   kubeadm |   1.22.6-00 | https://apt.kubernetes.io kubernetes-xenial/main arm64 Packages
   kubeadm |   1.22.5-00 | https://apt.kubernetes.io kubernetes-xenial/main arm64 Packages
   kubeadm |   1.22.4-00 | https://apt.kubernetes.io kubernetes-xenial/main arm64 Packages
   kubeadm |   1.22.3-00 | https://apt.kubernetes.io kubernetes-xenial/main arm64 Packages
   kubeadm |   1.22.2-00 | https://apt.kubernetes.io kubernetes-xenial/main arm64 Packages
   kubeadm |   1.22.1-00 | https://apt.kubernetes.io kubernetes-xenial/main arm64 Packages
   kubeadm |   1.22.0-00 | https://apt.kubernetes.io kubernetes-xenial/main arm64 Packages
```

Figure 3.2 – Available versions

Once we have made up our mind about which version we want to upgrade to, let's start prepping for the upgrade process:

1. We will start by upgrading kubeadm where we will need to use the following command to replace x in 1.23.x-00 with the latest patch version, which is 1.23.3 in our case:

```
apt-mark unhold kubeadm && \
apt-get update && apt-get install -y kubeadm=1.23.3-00 && \
apt-mark hold kubeadm
```

The output of the apt-mark command is the following:

```
kubeadm set on hold.
```

2. Now we can check the version of kubeadm with the `kubeadm version` command and see whether it's 1.23.3:

```
Client Version: version.Info{Major:"1",
Minor:"23", GitVersion:"v1.23.2",
GitCommit:"e6c093d87ea4cbb530a7b2ae91e54c0842d8308a",
GitTreeState:"clean", BuildDate:"2022-02-16T12:38:05Z",
GoVersion:"go1.17.7", Compiler:"gc", Platform:"linux/amd64"}
```

3. We use the `kubeadm upgrade plan` command to check whether the current cluster can be upgraded and the available versions that it can be upgraded to:

   `kubeadm upgrade plan`

   As shown in *Figure 3.3*, I can upgrade the kubelet and control plane components such as the API server, scheduler, and controller manager from 1.23.2 to 1.23.3:

```
Components that must be upgraded manually after you have upgraded the control plane with 'kubeadm upgrade apply':
COMPONENT    CURRENT        TARGET
kubelet      1 x v1.23.2    v1.23.3

Upgrade to the latest version in the v1.23 series:

COMPONENT                  CURRENT    TARGET
kube-apiserver             v1.23.2    v1.23.3
kube-controller-manager    v1.23.2    v1.23.3
kube-scheduler             v1.23.2    v1.23.3
kube-proxy                 v1.23.2    v1.23.3
CoreDNS                    v1.8.6     v1.8.6
etcd                       3.5.1-0    3.5.1-0
```

Figure 3.3 – kubeadm upgrade plan

4. If we decide to take action to upgrade the current cluster from 1.23.2 to 1.23.3, we can use the following command. Note that after `apply`, you just replace 1.23.3 for any future available versions that you wish to upgrade to:

   `kubeadm upgrade apply v1.23.3`

Important Note

To perform the upgrade operation smoothly, we recommend you get root permission in the exam by running the `sudo su` command.

In your daily upgrade task, you can use `sudo` and input your password to perform this operation.

Once you have given the command, you will get a message stating that the upgrade was a success:

```
upgrade/successful] SUCCESS! Your cluster was upgraded to "v1.23.3". Enjoy!
upgrade/kubelet] Now that your control plane is upgraded, please proceed with upgrading your kubelets if you haven't already done so.
```

Figure 3.4 – Control plane successfully upgraded

5. We then need to cordon the node, so we drain the workloads to prepare the node for maintenance. We cordon a node called `cloudmelonplaysrv` with the following command:

`kubectl drain cloudmelonplaysrv --ignore-daemonsets`

It will display a bunch of pods being evicted, which means those pods are being eliminated from the cordoned worker nodes:

```
node/cloudmelonplaysrv cordoned
WARNING: ignoring DaemonSet-managed Pods: calico-system/calico-node-xpvkh, kube-system/kube-proxy-75zqj
evicting pod tigera-operator/tigera-operator-768d489967-5xbth
evicting pod calico-system/calico-kube-controllers-7dddfdd6c9-sjdkf
evicting pod calico-apiserver/calico-apiserver-64cd47df68-hx2w7
evicting pod calico-apiserver/calico-apiserver-64cd47df68-j77cg
evicting pod kube-system/coredns-64897985d-57s4s
evicting pod calico-system/calico-typha-69f76d55b8-6xtwk
evicting pod kube-system/coredns-64897985d-712zx
pod/calico-typha-69f76d55b8-6xtwk evicted
pod/calico-kube-controllers-7dddfdd6c9-sjdkf evicted
pod/calico-apiserver-64cd47df68-j77cg evicted
pod/calico-apiserver-64cd47df68-hx2w7 evicted
pod/tigera-operator-768d489967-5xbth evicted
pod/coredns-64897985d-712zx evicted
pod/coredns-64897985d-57s4s evicted
node/cloudmelonplaysrv drained
```

Figure 3.5 – Draining workloads on the node

If you're using the `kubectl get no` command, the node will be marked as `schedulingDisabled`.

6. We use the following command to upgrade the kubelet and kubectl:

`apt-mark unhold kubelet kubectl && \`

`apt-get update && apt-get install -y kubelet=1.23.3-00 kubectl=1.23.3-00 && \`

`apt-mark hold kubelet kubectl`

7. Restart the kubelet:

`sudo systemctl daemon-reload`
`sudo systemctl restart kubelet`

8. Now we can uncordon the node and it will make the workloads schedulable again on the node that's being upgraded, called `cloudmelonplaysrv`:

`kubectl uncordon cloudmelonplaysrv`

This command will return the node that is now shown as `uncordoned`.

Upgrading the worker node

Since the worker node is where your workloads are actually up and running, we need to perform an upgrade one at a time and then replicate the same operation to all the other worker nodes available in the current Kubernetes cluster. *Figure 3.6* depicts the general upgrade workflow:

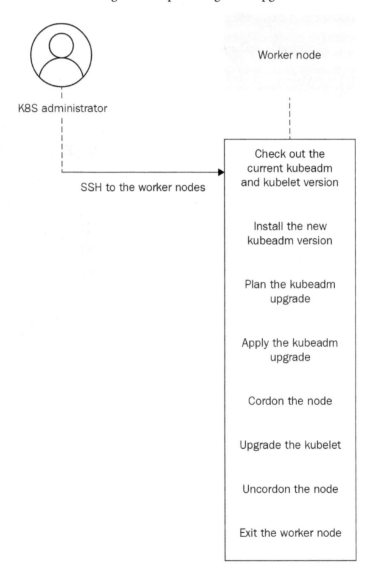

Figure 3.6 – Draining workloads on the node

1. Let's start with upgrading the kubeadm from 1.23.2 to 1.23.3 with the following command:

```
apt-mark unhold kubeadm && \
apt-get update && apt-get install -y kubeadm=1.23.3-00 && \
apt-mark hold kubeadm
```

2. For a worker node, we upgrade the kubelet, which also upgrades the local kubelet configuration, with the following command:

```
sudo kubeadm upgrade node
```

3. Similarly, we need to cordon the node so we drain the workloads to prepare the node for maintenance. Here, we are cordoning a node called `cloudmelonplayclient` using the following command:

```
kubectl drain cloudmelonplayclient --ignore-daemonsets
```

    We can then use the `kubectl get no` command to check the node status. It will be marked as `schedulingDisabled`.

4. We use the following command to upgrade the kubelet and kubectl just as we did for the master node:

```
apt-mark unhold kubelet kubectl && \
apt-get update && apt-get install -y kubelet=1.23.3-00
kubectl=1.23.3-00 && \
apt-mark hold kubelet kubectl
```

5. Restart the kubelet for the changes to take effect:

```
sudo systemctl daemon-reload
sudo systemctl restart kubelet
```

6. Finally, we can `uncordon` the node and it will make the workloads schedulable again on the node called `cloudmelonplayclient`. It will return the node that is now shown as uncordoned:

```
kubectl uncordon cloudmelonplayclient
```

We have now concluded the upgrade process for worker nodes. After the upgrade process, please make sure you use the `kubectl get nodes` command to make sure all the nodes have the `ready` status.

Working with etcd

Cluster data is stored in a key-value store in a Kubernetes cluster called etcd. The cluster data includes cluster state information such as pod state data, node state data, and configurations. As this data is critical for Kubernetes to orchestrate the workloads to the desired state, it stands to reason that it should be backed up periodically.

To access the etcd cluster inside the Kubernetes cluster, we can run the following command:

```
kubectl get po -n kube-system
```

This will list all the pods currently running in the `kube-system` namespace:

```
[cloudmelon@cloudmelonplaysrv:~$ kubectl get po -n kube-system
NAME                                             READY   STATUS    RESTARTS      AGE
coredns-64897985d-57s4s                          1/1     Running   0             14d
coredns-64897985d-7l2zx                          1/1     Running   0             14d
etcd-cloudmelonplaysrv                           1/1     Running   1             14d
kube-apiserver-cloudmelonplaysrv                 1/1     Running   4 (8d ago)    14d
kube-controller-manager-cloudmelonplaysrv        1/1     Running   4 (8d ago)    14d
kube-proxy-vjzv7                                 1/1     Running   0             14d
kube-scheduler-cloudmelonplaysrv                 1/1     Running   4 (8d ago)    14d
```

Figure 3.7 – Check the current etcd pod status

In the following sections, we'll take a closer look at the etcd cluster pod and learn all the related information that will be useful in the actual CKA exam.

Exploring the ETCD cluster pod

To get a closer look at the etcd pod that we have, use the following command:

```
kubectl describe po <etcd-podname> -n kube-system
```

For example, to get detailed information for an etcd pod called `etcd-cloudmelonplaysrv`, the command would be as follows:

```
kubectl describe po etcd-cloudmelonplaysrv -n kube-system
```

It returns the following output:

```
cloudmelon@cloudmelonplaysrv:~$ kubectl describe po etcd-cloudmelonplaysrv -n kube-system
Name:                 etcd-cloudmelonplaysrv
Namespace:            kube-system
Priority:             2000001000
Priority Class Name:  system-node-critical
Node:                 cloudmelonplaysrv/172.16.16.129
Start Time:           Tue, 25 Jan 2022 05:53:42 +0000
Labels:               component=etcd
                      tier=control-plane
Annotations:          kubeadm.kubernetes.io/etcd.advertise-client-urls: https://172.16.16.129:2379
                      kubernetes.io/config.hash: 449e5dd74342c261a2886f93dcda3eda
                      kubernetes.io/config.mirror: 449e5dd74342c261a2886f93dcda3eda
                      kubernetes.io/config.seen: 2022-01-25T05:53:41.313526419Z
                      kubernetes.io/config.source: file
                      seccomp.security.alpha.kubernetes.io/pod: runtime/default
Status:               Running
IP:                   172.16.16.129
IPs:
  IP:                 172.16.16.129
Controlled By:  Node/cloudmelonplaysrv
Containers:
  etcd:
    Container ID:   docker://84601b9c4fe85d81e0e861bc66bb6a3202a1ec1c8c450e939d221d0570db90bb
    Image:          k8s.gcr.io/etcd:3.5.1-0
    Image ID:       docker-pullable://k8s.gcr.io/etcd@sha256:64b9ea357325d5db9f8a723dcf503b5a449177b17ac87d69481e126bb724c263
    Port:           <none>
    Host Port:      <none>
    Command:
      etcd
      --advertise-client-urls=https://172.16.16.129:2379
      --cert-file=/etc/kubernetes/pki/etcd/server.crt
      --client-cert-auth=true
      --data-dir=/var/lib/etcd
      --initial-advertise-peer-urls=https://172.16.16.129:2380
      --initial-cluster=cloudmelonplaysrv=https://172.16.16.129:2380
      --key-file=/etc/kubernetes/pki/etcd/server.key
      --listen-client-urls=https://127.0.0.1:2379,https://172.16.16.129:2379
      --listen-metrics-urls=http://127.0.0.1:2381
      --listen-peer-urls=https://172.16.16.129:2380
      --name=cloudmelonplaysrv
      --peer-cert-file=/etc/kubernetes/pki/etcd/peer.crt
      --peer-client-cert-auth=true
      --peer-key-file=/etc/kubernetes/pki/etcd/peer.key
      --peer-trusted-ca-file=/etc/kubernetes/pki/etcd/ca.crt
      --snapshot-count=10000
      --trusted-ca-file=/etc/kubernetes/pki/etcd/ca.crt
    State:          Running
      Started:      Tue, 25 Jan 2022 05:53:36 +0000
    Ready:          True
    Restart Count:  1
    Requests:
      cpu:        100m
      memory:     100Mi
    Liveness:     http-get http://127.0.0.1:2381/health delay=10s timeout=15s period=10s #success=1 #failure=8
    Startup:      http-get http://127.0.0.1:2381/health delay=10s timeout=15s period=10s #success=1 #failure=24
    Environment:  <none>
    Mounts:
      /etc/kubernetes/pki/etcd from etcd-certs (rw)
      /var/lib/etcd from etcd-data (rw)
Conditions:
  Type              Status
  Initialized       True
  Ready             True
  ContainersReady   True
  PodScheduled      True
Volumes:
  etcd-certs:
    Type:          HostPath (bare host directory volume)
    Path:          /etc/kubernetes/pki/etcd
    HostPathType:  DirectoryOrCreate
  etcd-data:
    Type:          HostPath (bare host directory volume)
    Path:          /var/lib/etcd
    HostPathType:  DirectoryOrCreate
QoS Class:        Burstable
Node-Selectors:   <none>
Tolerations:      :NoExecute op=Exists
Events:           <none>
```

Figure 3.8 – Check the current etcd pod

In the figure, you can see the following important information about etcd:

```
etcd
    --advertise-client-urls=https://172.16.16.129:2379
    --cert-file=/etc/ubernetes/pki/etcd/server.crt
    --client-cert-auth=true
    --data-dir=/var/lib/etcd
    --initial-advertise-peer-urls=https://172.16.16.129:2380
    --initial-cluster=cloudmelonplaysrv=htt
ps://172.16.16.129:2380
    --key-file=/etc/ubernetes/pki/etcd/server.key
    --listen-client-urls=https://127.0.0.1:2379,htt
ps://172.16.16.129:2379
    --listen-metrics-urls=http://127.0.0.1:2381
    --listen-peer-urls=https://172.16.16.129:2380
    --name=cloudmelonplaysrv
    --peer-cert-file=/etc/ubernetes/pki/etcd/peer.crt
    --peer-client-cert-auth=true
    --peer-key-file=/etc/ubernetes/pki/etcd/peer.key
    --peer-trusted-ca-file=/etc/ubernetes/pki/etcd/ca.crt
    --snapshot-count=10000
    --trusted-ca-file=/etc/ubernetes/pki/etcd/ca.crt
  State:          Running
```

Among all the configurable parameters, the following will come in handy in your daily work with etcd:

- `--advertise-client-urls` tells etcd to accept incoming requests from the clients. It accepts a list of URLs.

- `--cert-file` is where we specify the client server TLS `cert` file path.

- `--key-file` is where we specify the client server TLS `key` file path.

- `--trusted-ca-file` is where we specify the client server TLS `trusted CA cert` file path.

These are key flags that will authenticate your request from the client with secure communication. You will need them to check the etcd status, backup, and restore etcd cluster.

> **Important Note**
>
> Access to etcd is equivalent to getting root permission in the cluster. We make sure the authentication request is only going through the API server.

To know more about other configurable parameters, please check out `https://etcd.io/docs/v3.5/op-guide/configuration/`.

Listing etcd cluster members

With the information that we acquired from the `kubectl describe pod` command, we can list the members of the etcd cluster:

```
kubectl exec etcd-cloudmelonplaysrv -n kube-system --
sh -c "ETCDCTL_API=3 etcdctl member list --endpoints=ht
tps://127.0.0.1:2379 --cacert=/etc/kubernetes/pki/etcd/
ca.crt --cert=/etc/kubernetes/pki/etcd/server.crt --key=/etc/
kubernetes/pki/etcd/server.key"
```

It returns the information about members. In our case, we have only one result because we are working with a single master node. Our command will look like the following:

```
8d1f17827821818f, started, cloudmelonplaysrv,
https://172.16.16.129:2380, https://172.16.16.129:2379, false
```

The output describes columns such as `ID` and `Status` of the etcd cluster, the etcd cluster name, and the peer and client address.

You can form the output automatically in tabular form with `--write-out=table`. It will look like this:

ID	STATUS	NAME	PEER ADDRS	CLIENT ADDRS	IS LEARNER
8d1f17827821818f	started	cloudmelonplaysrv	https://172.16.16.129:2380	https://172.16.16.129:2379	false

Figure 3.9 – The current etcd member list

Notice that the client address is the same as the value of the `--advertise-client-urls` URL in the output of `kubectl describe pod`.

Checking the etcd cluster status

You can use the following command to check the etcd cluster status and write the output in tabular form. Note that using the correct etcdctl API version, we're on API version 3 in the following example:

```
ETCDCTL_API=3 etcdctl endpoint status
```

The following command is used to access an etcd pod from the Kubernetes cluster and check out the status of the etcd pod in the multi-node etcd cluster:

```
kubectl -n kube-system exec <etcd-podname> -- sh -c "ETCDCTL_
API=3 etcdctl endpoint status --write-out=table --endpoint
s=https://<IP1>:2379,https://<IP2>:2379,https://<IP3>:2379
--cacert=/etc/kubernetes/pki/etcd/ca.crt --cert=/etc/
kubernetes/pki/etcd/server.crt --key=/etc/kubernetes/pki/etcd/
server.key"
```

You can use the information that you acquired from etcdctl member list in this command:

- ETCDCTL_API is the etcdctl version.

- - - endpoints are the client addresses of your etcd members if you have multiple master nodes.

In this chapter, however, we are showing off a single master node and it contains only one etcd member. Therefore, this command to access an etcd pod called etcd-cloudmelonplaysrv from the Kubernetes cluster and check out the status of the etcd pod will look like this:

```
kubectl -n kube-system exec etcd-cloudmelonplaysrv -- sh
-c "ETCDCTL_API=3 etcdctl endpoint status --endpoints=ht
tps://172.16.16.129:2379 --cacert=/etc/kubernetes/pki/etcd/
ca.crt --cert=/etc/kubernetes/pki/etcd/server.crt --key=/etc/
kubernetes/pki/etcd/server.key --write-out=table"
```

It will look like the following output:

ENDPOINT	ID	VERSION	DB SIZE	IS LEADER	IS LEARNER	RAFT TERM	RAFT INDEX	RAFT APPLIED INDEX	ERRORS
https://172.16.16.129:2379	8d1f17827821818f	3.5.1	5.1 MB	true	false	2	2403920	2403920	

Figure 3.10 – The current etcd member list

From the output of kubectl describe pod < etcd-podname>, we also learn that we have two listen client IPs:

```
--listen-client-urls=https://127.0.0.1:2379,ht
tps://172.16.16.129:2379
```

As we're checking the etcd cluster status inside the Kubernetes cluster, we can also use the internal endpoint address `https://127.0.0.1:2379` to check the etcd cluster status. The following command can be used to access an etcd pod named `etcd-cloudmelonplaysrv` from the Kubernetes cluster and check out the status of the etcd pod with the internal endpoint:

```
kubectl -n kube-system exec etcd-cloudmelonplaysrv -- sh
-c "ETCDCTL_API=3 etcdctl endpoint status --endpoints=h
ttps://127.0.0.1:2379 --cacert=/etc/kubernetes/pki/etcd/
ca.crt --cert=/etc/kubernetes/pki/etcd/server.crt --key=/etc/
kubernetes/pki/etcd/server.key --write-out=table"
```

And it returns the information regarding the etcd cluster:

ENDPOINT	ID	VERSION	DB SIZE	IS LEADER	IS LEARNER	RAFT TERM	RAFT INDEX	RAFT APPLIED INDEX	ERRORS
https://127.0.0.1:2379	8d1f17827821818f	3.5.1	5.1 MB	true	false	2	2404458	2404458	

Figure 3.11 – The current etcd member list

In the following section, we'll explore interacting with the etcd cluster from the client outside of the Kubernetes cluster.

Installing etcd

To access etcd outside of the Kubernetes cluster, you will need to install etcdctl. You can do so by following the instructions in this section. Please note, however, that this is not part of the CKA exam.

To get started, we'll need to get the etcd binary:

```
wget https://github.com/etcd-io/etcd/releases/download/
v3.4.18/etcd-v3.4.18-linux-amd64.tar.gz
tar xvf etcd-v3.4.18-linux-amd64.tar.gz
sudo mv etcd-v3.4.18-linux-amd64/etcd* /usr/local/bin
```

Once you finish the installation, you can use the following command to verify the current version:

```
etcdctl version
```

The command returns the current etcdctl client version and the API version in the following manner:

```
[cloudmelon@cloudmelonplaysrv:~$ etcdctl version
etcdctl version: 3.5.0
API version: 3.5
```

Figure 3.12 – Check the current etcdctl version outside of the Kubernetes cluster

Similarly, you can use the following command to check the kubectl version in the Kubernetes cluster. When you're using the `kubectl exec` command, it executes directly on the pod named `etcd-cloudmelonplaysrv` located in the `kube-system` namespace. We can use the following command to execute the `etcdctl version` Bash command to get the version of the etcd store:

```
kubectl exec etcd-cloudmelonplaysrv -n kube-system -- sh -c
"etcdctl version"
```

The returned result is similar to the etcdctl client version and the API version:

```
[cloudmelon@cloudmelonplaysrv:~$ kubectl exec etcd-cloudmelonplaysrv -n kube-system -- sh -c "etcdctl version"
etcdctl version: 3.5.1
API version: 3.5
```

Figure 3.13 – Check the current etcdctl version in the Kubernetes cluster

Similarly, once you have etcdctl installed, you can check the etcd store status by running the following command and you'll get the endpoint status:

```
kubectl exec etcd-cloudmelonplaysrv -n kube-system -- sh
-c " etcdctl --write-out=table --endpoints=$ENDPOINTS endpoint
status "
```

```
If you want to make sure the etcd store is healthy, using the
following command with the endpoint from the Previous command :
kubectl exec etcd-cloudmelonplaysrv -n kube-system -- sh -c
" etcdctl --endpoints=$ENDPOINTS endpoint health "
```

Backing up etcd

With all the groundwork we have laid out in the previous sections, we can generalize the backup etcd process as follows:

1. SSH to the etcd cluster node. It could be a separate node or the same as the master node. In the CKA exam, it's likely you'll be starting in the master node, where etcdctl is installed; thus, this step is optional.

2. Check out the etcd status. You could acquire the necessary information from the `kubectl describe <etcd-podname>` command.

3. Perform the etcd backup.

4. Exit the master node. This may not be necessary in the actual CKA exam.

The general process is captured in the following diagram:

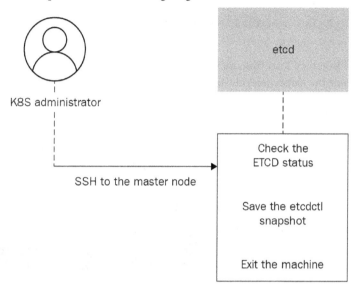

Figure 3.14 – Backup etcd process

Now let's look at the detailed process of how to back up etcd:

1. If you need to connect to the master node or the etcd cluster node, you can use the `ssh master-0` command or the `ssh username@<nodeIP>` command. Please note that this step is optional. Following is an example of a user named `packtuser` using `ssh` to connect to a node with the IP address `10.10.11.20`:

    ssh packtuser@10.10.11.20

2. Check the etcd status using the following command from outside the cluster:

    **sudo ETCDCTL_API=3 etcdctl endpoint status --endpoints=ht
    tps://172.16.16.129:2379 --cacert=/etc/kubernetes/pki/etcd/
    ca.crt --cert=/etc/kubernetes/pki/etcd/server.crt --key=/etc/
    kubernetes/pki/etcd/server.key --write-out=table**

The output returned will be as follows:

ENDPOINT	ID	VERSION	DB SIZE	IS LEADER	IS LEARNER	RAFT TERM	RAFT INDEX	RAFT APPLIED INDEX	ERRORS
https://172.16.16.129:2379	8d1f17827821818f	3.5.1	5.1 MB	true	false	2	2452141	2452141	

Figure 3.15 – Check the etcd status from outside of the cluster

3. Back up the etcd cluster using the `etcdctl snapshot save` command. It will look like this:

```
sudo ETCDCTL_API=3 etcdctl --endpoints $ENDPOINT snapshot save
snapshotdb
```

You will need to authenticate from the API server with secure communication as you're backing up from outside the Kubernetes cluster. For this, you can use the following command:

```
sudo ETCDCTL_API=3 etcdctl snapshot save snapshotdb
--endpoints=https://172.16.16.129:2379
--cacert=/etc/kubernetes/pki/etcd/ca.crt --cert=/etc/
kubernetes/pki/etcd/server.crt --key=/etc/kubernetes/pki/etcd/
server.key
```

The returned output shows that you have backed up the etcd store successfully:

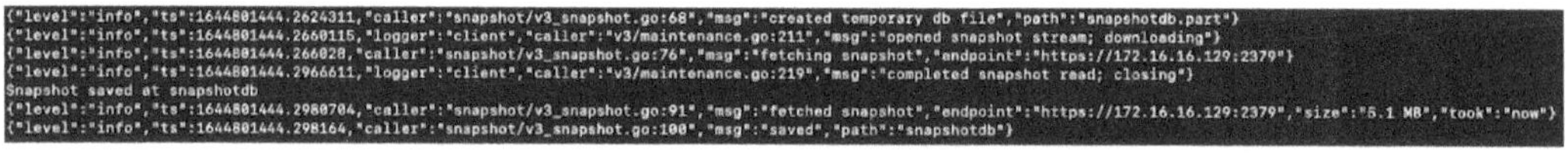

{"level":"info","ts":1644801444.2624311,"caller":"snapshot/v3_snapshot.go:68","msg":"created temporary db file","path":"snapshotdb.part"}
{"level":"info","ts":1644801444.2660115,"logger":"client","caller":"v3/maintenance.go:211","msg":"opened snapshot stream; downloading"}
{"level":"info","ts":1644801444.266028,"caller":"snapshot/v3_snapshot.go:76","msg":"fetching snapshot","endpoint":"https://172.16.16.129:2379"}
{"level":"info","ts":1644801444.2966611,"logger":"client","caller":"v3/maintenance.go:219","msg":"completed snapshot read; closing"}
Snapshot saved at snapshotdb
{"level":"info","ts":1644801444.2988784,"caller":"snapshot/v3_snapshot.go:91","msg":"fetched snapshot","endpoint":"https://172.16.16.129:2379","size":"5.1 MB","took":"now"}
{"level":"info","ts":1644801444.298164,"caller":"snapshot/v3_snapshot.go:100","msg":"saved","path":"snapshotdb"}

Figure 3.16 – Backup etcd store

4. Verify the snapshot via the following command:

```
sudo ETCDCTL_API=3 etcdctl snapshot status snapshotdb --endpo
ints=https://172.16.16.129:2379 --cacert=/etc/kubernetes/pki/
etcd/ca.crt --cert=/etc/kubernetes/pki/etcd/server.crt --key=/
etc/kubernetes/pki/etcd/server.key --write-out=table
```

The following figure shows the status of the etcd cluster:

ENDPOINT	ID	VERSION	DB SIZE	IS LEADER	IS LEARNER	RAFT TERM	RAFT INDEX	RAFT APPLIED INDEX	ERRORS
https://172.16.16.129:2379	8d1f17827821818f	3.5.1	5.1 MB	true	false	2	2455079	2455079	

Figure 3.17 – Check the etcd store with the snapshot backup

Restoring etcd

To restore etcd clusters, you can follow the process depicted in *Figure 3.18*. Note that if you have any API server instances running, you need to stop them before performing the restore operation. You may restart the API server instances after the etcd is restored:

1. SSH to the etcd cluster node.
2. Check the etcd status.

3. Restore the etcd backup.

4. Exit the master node:

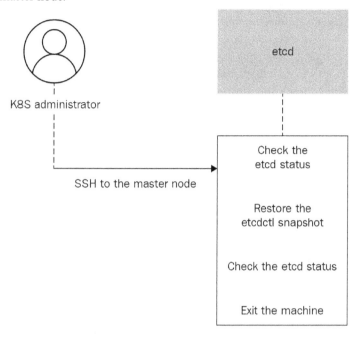

Figure 3.18 – Restore etcd process

Once you have a snapshot present, you can restore etcd from the previous backup operation using the following command:

```
sudo ETCDCTL_API=3 etcdctl --endpoints 172.16.16.129:2379
snapshot restore snapshotdb
```

The returned output shows that the etcd store has been restored successfully:

Figure 3.19 – Restored etcd store with an existing snapshot

You have now completed the etcd restore process.

Note that this approach can be used if you want to restore the etcd cluster from a different patch version. It is important to back up etcd regularly, then perform the restore operation to recover the cluster data from a failed cluster. To learn more about etcd cluster backup and restore for Kubernetes, please check out https://kubernetes.io/docs/tasks/administer-cluster/configure-upgrade-etcd/.

Summary

This chapter covers one of the most common jobs of a Kubernetes administrator – that is, maintaining and upgrading Kubernetes clusters. Similar to cluster installation, this is also one of the most time-consuming tasks in the CKA exam. Again, practice makes perfect. The HA topology for a Kubernetes cluster in *Chapter 2, Installing and Configuring Kubernetes Cluster,* helps you understand what you are going to upgrade and how to do it. If needed, go back to *Chapter 1, Kubernetes Overview*, and make sure that you have a good understanding of the Kubernetes components. This way, you will know how and what's needed to upgrade the control plane and worker nodes.

Compared to cluster upgrades, backup and restore etcd is one of the best-in-value questions in the CKA exam as it is simple to answer with a high-value score. Thoroughly practicing what we've learned in this chapter will help you overcome any challenges in the exam.

In the next chapter, we'll talk about application scheduling and life cycle management, where we will revisit some important Kubernetes objects and concepts, and touch upon how they play out both in the CKA exam and in real life. Stay tuned!

Mock CKA scenario-based practice test

You have two virtual machines, master-0 and worker-0. Please complete the following mock scenarios:

Scenario 1

SSH to the master-0 node, check the current kubeadm version, and upgrade to the latest kubeadm version. Check out the current kubectl version, and upgrade to the latest kubectl version.

Scenario 2

SSH to worker-0 node, check out the current kubeadm version, and upgrade to the latest kubeadm version. Check out the current kubelet version, and upgrade to the latest kubelet version.

Scenario 3

SSH to the master-0 node and back up the etcd store.

Scenario 4

SSH to the master-0 node and restore the etcd store to the previous backup.

You can find all the scenario resolutions in *Appendix - Mock CKA scenario-based practice test resolutions* of this book.

FAQs

1. *Where can I find out about the compatible version of Kubernetes components with each release?*

   Go to the Kubernetes official documentation to learn about the version skew policy: `https://kubernetes.io/releases/version-skew-policy/`.

2. *Where can I learn about the latest developments of the etcd store?*

   Go to `https://etcd.io/`, where you will find the latest developments of the etcd store. For daemons and guidance on how to get started with etcd, please go to the official documentation: `https://etcd.io/docs/`.

3. *What is a recommended official Kubernetes article for upgrading a Kubernetes cluster?*

   I recommend bookmarking the article *Upgrading the kubeadm*, where you will find most key commands and processes: `https://kubernetes.io/docs/tasks/administer-cluster/kubeadm/kubeadm-upgrade/`.

4. *What is a recommended official Kubernetes article for backup and restore etcd?*

   I recommend bookmarking the article *Operating etcd clusters for Kubernetes*, where you will find all the key commands for etcd backup and restore: `https://kubernetes.io/docs/tasks/administer-cluster/configure-upgrade-etcd/`.

Part 2: Managing Kubernetes

This part describes how to manage workloads deployed on top of Kubernetes, and how to manage the security and networking of Kubernetes clusters to fulfil enterprise requirements.

This part of the book comprises the following chapters:

4

Application Scheduling and Lifecycle Management

This chapter describes how to use Kubernetes deployments to deploy pods, scale pods, perform rolling updates and rollbacks, carry out resource management, and use ConfigMaps to configure pods using `kubectl` commands and YAML definitions. This chapter covers 15% of the CKA exam content.

In this chapter, we're going to cover the following main topics:

- The basics of Kubernetes workloads
- Deploying and managing applications
- Scaling applications
- Performing rolling updates and rollbacks
- Resource management
- Workload scheduling
- Configuring applications

Technical requirements

To get started, we need to make sure your local machine meets the following technical requirements:

- A compatible Linux host – we recommend a Debian-based Linux distribution such as Ubuntu 18.04 or later
- Make sure your host machine has at least 2 GB RAM, 2 CPU cores, and about 20 GB of free disk space

The basics of Kubernetes workloads

Kubernetes orchestrates your workloads to achieve the desired status – a containerized workload with applications running on Kubernetes, including stateless, stateful, and data-processing applications. In terms of cloud-native applications, there's an interesting white paper that introduced the notion of cloud-native applications and design patterns thoroughly, which you can check out here if you're interested: `https://www.redhat.com/en/resources/cloud-native-container-design-whitepaper`.

The fundamental building blocks of any containerized workload up and running in the Kubernetes cluster are called Kubernetes API primitives or Kubernetes objects. They are the API resource types defined in Kubernetes, including pods, ReplicaSets, DaemonSets, StatefulSets, Job and CronJob objects, and Deployments, among others mentioned in *Chapter 1, Kubernetes Overview*.

The CKA exam covers some of the main Kubernetes objects such as Pods, Deployments, ReplicaSets, and DaemonSets while working with Kubernetes clusters and we'll dive into further detail in the following section of this chapter.

Please make sure your local machine meets the required technical requirements before diving into the practice.

Imperative management versus declarative management

There are a few ways to communicate with API servers in Kubernetes – mainly, they can be categorized as either imperative management or declarative management. You will need to use both `kubectl` and YAML definitions to manage Kubernetes objects. The `kubectl` utilities can support all the management techniques for managing Kubernetes objects, as Kubernetes is intended to be a desired state manager. After executing a `kubectl` command, as a result, it moves the current workload running in Kubernetes from its actual state to the desired state, which is defined in the command-line parameters or YAML-defined specifications.

Time management is the key to success in the CKA exam. Getting familiar with `kubectl` commands will help you save a lot of time when it comes to a new deployment. A good understanding of YAML definition will help you update the configurations quickly.

Understanding pods

The smallest deployable unit in Kubernetes is a pod. The pod contains the actual application workload – it could be one or multiple containers. A pod in Kubernetes has a defined lifecycle. We'll cover the following topic about pods:

- Understanding pods
- Understanding health probing for pods

- Understanding a multi-container pod
- Understanding an init container
- Understanding a static pod

Let's take a look at the pod first. You can create a pod using an imperative command as follows:

```
kubectl run <pod-name> --image=<image-name:image-tag>
```

This is an example of running a pod named `ngin-pod` with the image as `nginx` and the image tag as `alpine`:

```
kubectl run nginx-pod --image=nginx:alpine
```

You will see the output is returned as `created`, as follows, to indicate that your pod has been created successfully:

```
pod/nginx-pod created
```

In the process, you will see pod has the `ContainerCreating` status, indicating that the container is being created, and you can use `kubectl` to describe a pod command to see what's going on. The following command is what we can use to check the pod's current status:

```
kubectl describe pod nginx-pod
```

At the bottom of the `describe` command, you will see the events – this is helpful information for you to use to check whether anything is going wrong during your deployment. We will explore troubleshooting pods further in *Chapter 8, Monitoring and Logging Kubernetes Clusters and Applications*:

```
Events:
  Type    Reason     Age   From               Message
  ----    ------     ----  ----               -------
  Normal  Scheduled  98s   default-scheduler  Successfully assigned default/nginx-pod to cloudmelonplayground
  Normal  Pulling    97s   kubelet            Pulling image "nginx:alpine"
  Normal  Pulled     11s   kubelet            Successfully pulled image "nginx:alpine" in 1m25.997135506s
  Normal  Created    11s   kubelet            Created container nginx-pod
  Normal  Started    11s   kubelet            Started container nginx-pod
```

Figure 4.1 – The pod events

The same pod can be YAML-defined, as follows, which will give you the same result:

```
apiVersion: v1
kind: Pod
metadata:
  name: nginx
spec:
```

```
containers:
- name: nginx
  image: nginx:alpine
  ports:
  - containerPort: 80
```

You can use the following command to deploy a YAML definition:

```
kubectl apply -f <your-spec>.yaml
```

Similarly, we can run a BusyBox image with a single command, such as the following:

```
kubectl run busybox --rm -it --image=busybox /bin/sh
```

You can also deploy a nginx image and then export the YAML definition by using the -o yaml flag:

```
kubectl run nginx --image=nginx --dry-run -o yaml > pod-sample.
yaml
```

After running this command, a sample yaml file will be exported to your local PC – you can edit this yaml file to make changes locally if needed.

Understanding liveness, readiness, and startup probes

To explore the health status of the pods further, let's talk about health probes. Probes allow you to know how Kubernetes determines the states of your containers. Let's have look at each of them one by one:

- **Liveness probes** indicate whether the container is running properly, as they govern when the cluster will decide to restart the container automatically.

- **Readiness probes** indicate whether the container is ready to accept requests.

- **Startup probes** check when a container starts and are very handy for containers that require an additional startup time on their first initialization, preventing them from being killed by kubelet before they get on their feet. Once configured, they disable liveness and readiness checkers until they're complete.

We'll have a look at these in more detail in *Chapter 8, Monitoring and Logging Kubernetes Clusters and Applications*. You can find further details about health probes at the following link: https://kubernetes.io/docs/tasks/configure-pod-container/configure-liveness-readiness-startup-probes/.

Understanding a multi-container pod

Multi-container pods are simply pods with more than one container working together as a single unit. When it comes to multiple containers residing in a pod, a container interacts with another in the following two ways:

- **Shared networking**: When two containers are running on the same host when they are in the same pod, they can access each other by simply using *localhost*. All the listening ports are accessible to other containers in the pod, even if they're not exposed outside the pod.

  Figure 4.2 shows how multiple containers in the same pod share a local network with each other:

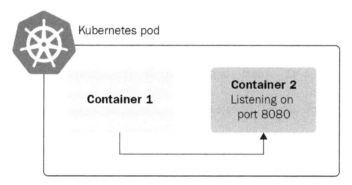

Figure 4.2 – A multi-container pod's shared network

- **Shared storage volumes**: We can mount the same volume to two different containers so that they can both interact with the same data – it is possible to have one container write data to the volume and the other container read that data from the same volume. Some volumes even allow concurrent reading and writing. We'll dive deeper into how storage works for multi-container pods in *Chapter 5, Demystifying Kubernetes Storage*.

Figure 4.3 shows how multiple containers in the same pod share local storage with each other:

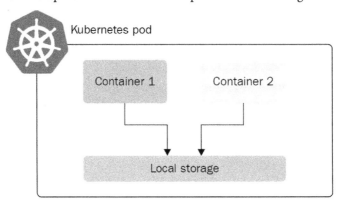

Figure 4.3 – A multi-container pod's shared storage volume

The following is an example of how to create multiple containers in a pod:

```
apiVersion: v1
kind: Pod
metadata:
  name: multi-app-pod
  labels:
      app: multi-app
spec:
  containers:
  - name: nginx
    image: nginx
    ports:
    - containerPort: 80
  - name: busybox-sidecar
    image: busybox
    command: ['sh', '-c', 'while true; do sleep 3600; done;']
```

In general, it is good to have a one-to-one relationship between a container and a pod, which follows the principles of building microservices by keeping each module independent. The real world is sometimes more complicated than it may seem, let's take a look at multi-container pods.

Understanding an init container

An init container is configured in a pod to execute before the container host starts. It is specified inside an `initContainers` section, as in the following example. You can configure multiple init containers too, which will allow each init container to complete one at a time in sequential order:

```
apiVersion: v1
kind: Pod
metadata:
  name: melon-pod
  labels:
    app: melonapp
spec:
  containers:
  - name: melonapp-container
    image: busybox:latest
    command: ['sh', '-c', 'echo The melonapp is running! &&
sleep 3600']
  initContainers:
  - name: init-melonservice
    image: busybox:latest
    command: ['sh', '-c', 'until nslookup melonservice; do echo
waiting for melonservice; sleep 2; done;']
  - name: init-melondb
    image: busybox:latest
    command: ['sh', '-c', 'until nslookup melondb; do echo
waiting for melondb; sleep 2; done;']
```

In the case that any of the init containers fail to complete, Kubernetes will restart the pod repeatedly until the init container succeeds. To learn more about init containers, visit the following link: https://kubernetes.io/docs/concepts/workloads/pods/init-containers/.

Understanding a static Pod

As the captain of a worker node, the `kubelet` agent can manage a node independently, and it can create pods. The pods that are managed directly by the `kubelet` daemon and bound to a specific node are called static pods. As opposed to pods that are managed by the Kubernetes master, static pods are watched by the `kubelet` agent, and it restarts in the case of failure.

The way to configure `kubelet` so that it reads the pod definition files is to add a YAML specification under the following directory where the static pod information is stored:

```
/etc/kubernetes/manifests
```

`kubelet` checks this directory periodically. This path can be configured in `kubelet.service`.

Understanding Job and CronJob objects

Jobs can be used to reliably execute a workload and define when it completes – typically, a Job will create one or more pods. After the Job is finished, the containers will exit and the pods will enter the `Completed` status.

Jobs can be used to reliably execute a workload until it completes. The Job will create one or more pods. When the Job is finished, the containers will exit and the pods will enter the `Completed` status. An example use of Jobs is when we want to run a particular workload and make sure that it runs once and succeeds.

1. You can create a Job with a YAML description:

    ```yaml
    apiVersion: batch/v1
    kind: Job
    metadata:
      name: pi
    spec:
      template:
        spec:
          containers:
          - name: pi
            image: perl
            command: ["perl",  "-Mbignum=bpi", "-wle", "print bpi(2000)"]
          restartPolicy: Never
      backoffLimit: 4
    ```

The backoffLimit parameter means that, if it fails 4 times, this is the limit. All the Job does the same as it is while creating a pod under the hood. Although a normal pod is constantly running, when a Job is complete, it goes into the Completed status. This means that the container is no longer running, so the pod still exists, but the container is complete.

2. You can use the following command to deploy a YAML definition:

```
kubectl apply -f melon-job.yaml
```

3. You can run the following command to check the Job's status:

```
kubectl get job
```

4. When the Job is still running, you can see the Running status. When the Job is finished, you can see that it is complete from the following:

```
[root@cloudmelonplayground:~# k get job
NAME    COMPLETIONS    DURATION    AGE
pi      1/1            27s         2m27s
```

Figure 4.4 – The Job is complete

CronJobs, based on the capability of a Job, add value by allowing users to execute Jobs on a schedule. Users can use cron expressions to define a particular schedule as per their requirements. The following is an example of a CronJob YAML definition:

```
apiVersion: batch/v1
kind: CronJob
metadata:
 name: hello
spec:
 schedule: "*/1 * * * *"
 jobTemplate:
   spec:
     template:
       spec:
         containers:
         - name: hello
           image: busybox
```

```
        args:
        - /bin/sh
        - -c
        - date; echo Hello from the Kubernetes cluster
      restartPolicy: OnFailure
```

5. You can use the following command to deploy a YAML definition:

kubectl apply -f melon-cronjob.yaml

You can use the following command to check the cron job's status:

kubectl get cronjob

You'll get an output as follows:

```
[root@cloudmelonplayground:~# k get cronjob
NAME     SCHEDULE      SUSPEND    ACTIVE    LAST SCHEDULE    AGE
hello    */1 * * * *   False      0         8s               79s
```

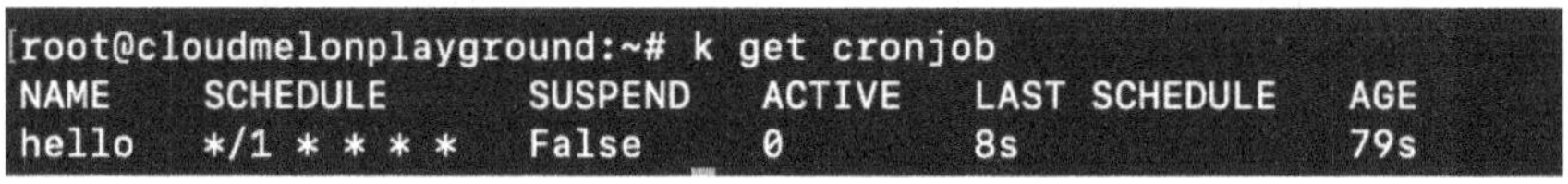

Figure 4.5 – The cron job shown as complete

This cron job creates a few pods name `hello`, so we will use the following command to check the log of the Job:

kubectl get pods | grep hello

You'll get an output as follows:

```
hello-27435182-1d9gw    0/1    Completed    0    2m26s
hello-27435183-8ptg8    0/1    Completed    0    86s
hello-27435184-htsxd    0/1    Completed    0    26s
```

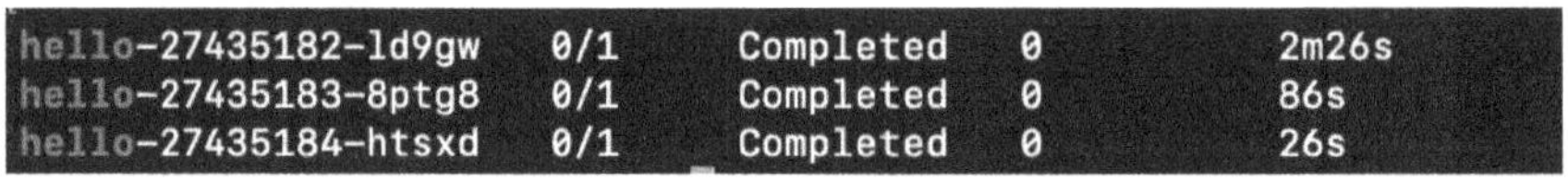

Figure 4.6 – The completed cron job pods

We can check the logs of these pods with the following command:

```
kubectl logs hello-xxxx
```

We can see that the cron job has been executed:

```
root@cloudmelonplayground:~# k logs hello-27435184-htsxd
Tue Mar  1 05:04:07 UTC 2022
Hello from the Kubernetes cluster
```

Figure 4.7 – The logs showing how the cron job was completed

If you want to delete cron jobs, you can use the following command:

```
kubectl delete cronjobs hello
```

Then, you will see the following output indicating that your cron job has been deleted:

```
cronjob.batch "hello" deleted
```

CronJobs were promoted to general availability in Kubernetes v1.21. You can find a great article about running automated tasks using a CronJob here: https://kubernetes.io/docs/tasks/job/automated-tasks-with-cron-jobs.

Deploying and managing applications

The following sections of this chapter will take you through practical exercises with concrete examples that you would encounter in your real CKA exam, including how to deploy and scale applications, perform rolling updates and rollbacks for those applications, manage and govern the resource consumption for these applications, and configure them.

Deploying applications

Deploying applications can be achieved in various ways, such as deploying a pod with kubectl or a YAML definition, as we did in the *The basics of Kubernetes workloads* section of this chapter. Now, we'll take a look at a more effective way of using Deployments. In this section, let's get into how to deploy and scale applications.

Deployments

A Deployment is a convenient way to define the desired state deployment – it provides us with a better way of upgrading the underlying instances seamlessly using rolling updates, undoing changes, and pausing and resuming changes as required. For example, things such as deploying a ReplicaSet with a certain number of replicas are easy to roll out and roll back, and more effective. The following figure depicts how a Deployment looks conceptually:

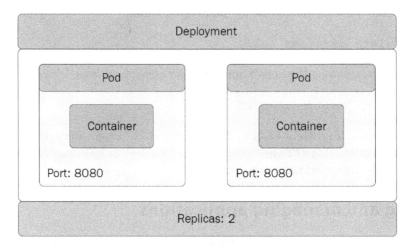

Figure 4.8 – A Deployment

Deployments provide a way to define a desired state for the replica pod. You can use a YAML definition as follows to define a Deployment:

```
apiVersion: apps/v1
kind: Deployment
metadata:
  name: nginx-deployment
  labels:
    app: nginx
spec:
  replicas: 3
  selector:
    matchLabels:
      app: nginx
  template:
    metadata:
```

```
    labels:
      app: nginx
  spec:
    containers:
    - name: nginx
      image: nginx:latest
      ports:
      - containerPort: 80
```

The following attributes are important to help you understand the preceding YAML definition:

- spec.replicas gives us the number of replica pods

- spec.template is the template pod descriptor that defines the pods that will be created

- spec.selector is the deployment that will manage all pods whose labels match this selector

We can create a Deployment using the following kubectl command:

```
kubectl create deployment kubeserve --image=nginx:latest
```

After running the preceding command, you will then get the following output:

```
deployment.apps/kubeserve created
```

You can use kubectl get deploy to query all the Deployments in the current namespace as follows:

```
kubectl get deployments
```

You will see the following Deployment status in the output:

```
[root@cloudmelonplayground:~# k get deployment kubeserve
NAME          READY    UP-TO-DATE    AVAILABLE    AGE
kubeserve     0/1      1             0            15s
```

Figure 4.9 – kubectl getting the Deployments

If you know the name of a Deployment, you can use the following command to get that Deployment:

```
kubectl get deployment kubeserve
```

You will see the following output:

```
[^Croot@cloudmelonplayground:~# k get deploy kubeserve
NAME        READY   UP-TO-DATE   AVAILABLE   AGE
kubeserve   1/1     1            1           2m3s
```

Figure 4.10 – kubectl getting a Deployment by name

The following command allows you to get the details of the Deployment:

```
kubectl describe deployment kubeserve
```

This command will help you understand the configurations in the Deployment, where you will see the following output:

```
root@cloudmelonplayground:~# k describe deploy kubeserve
Name:                   kubeserve
Namespace:              default
CreationTimestamp:      Wed, 02 Mar 2022 04:27:12 +0000
Labels:                 app=kubeserve
Annotations:            deployment.kubernetes.io/revision: 1
Selector:               app=kubeserve
Replicas:               1 desired | 1 updated | 1 total | 1 available | 0 unavailable
StrategyType:           RollingUpdate
MinReadySeconds:        0
RollingUpdateStrategy:  25% max unavailable, 25% max surge
Pod Template:
  Labels:  app=kubeserve
  Containers:
   nginx:
    Image:        nginx:latest
    Port:         <none>
    Host Port:    <none>
    Environment:  <none>
    Mounts:       <none>
  Volumes:        <none>
Conditions:
  Type           Status  Reason
  ----           ------  ------
  Available      True    MinimumReplicasAvailable
  Progressing    True    NewReplicaSetAvailable
OldReplicaSets:  <none>
NewReplicaSet:   kubeserve-dfbbf9445 (1/1 replicas created)
Events:
  Type    Reason           Age    From                   Message
  ----    ------           ----   ----                   -------
  Normal  ScalingReplicaSet  2m12s  deployment-controller  Scaled up replica set kubeserve-dfbbf9445 to 1
```

Figure 4.11 – kubectl describing a Deployment

The following command allows you to live-edit the Deployments:

```
kubectl edit deployment kubeserve
```

The preceding command is a magical one that will allow you to *live-edit* a Deployment. The following is the sample output and you can edit it live – it works similarly to when you create a pod using the vim editor. You can live-edit the Deployment here, and then save and quit using wq!:

```
# Please edit the object below. Lines beginning with a '#' will be ignored,
# and an empty file will abort the edit. If an error occurs while saving this file will be
# reopened with the relevant failures.
#
apiVersion: apps/v1
kind: Deployment
metadata:
  annotations:
    deployment.kubernetes.io/revision: "1"
  creationTimestamp: "2022-03-02T05:14:42Z"
  generation: 1
  labels:
    app: kubeserve
  name: kubeserve
  namespace: default
  resourceVersion: "198806"
  uid: f61d4c4a-20a6-42f6-94b9-2c974a551eb6
spec:
  progressDeadlineSeconds: 600
  replicas: 1
  revisionHistoryLimit: 10
  selector:
    matchLabels:
      app: kubeserve
  strategy:
    rollingUpdate:
      maxSurge: 25%
      maxUnavailable: 25%
    type: RollingUpdate
  template:
    metadata:
      creationTimestamp: null
      labels:
        app: kubeserve
    spec:
      containers:
      - image: nginx:latest
        imagePullPolicy: Always
        name: nginx
        resources: {}
        terminationMessagePath: /dev/termination-log
        terminationMessagePolicy: File
      dnsPolicy: ClusterFirst
      restartPolicy: Always
      schedulerName: default-scheduler
      securityContext: {}
      terminationGracePeriodSeconds: 30
status:
  availableReplicas: 1
  conditions:
  - lastTransitionTime: "2022-03-02T05:15:17Z"
    lastUpdateTime: "2022-03-02T05:15:17Z"
    message: Deployment has minimum availability.
    reason: MinimumReplicasAvailable
    status: "True"
    type: Available
  - lastTransitionTime: "2022-03-02T05:14:42Z"
    lastUpdateTime: "2022-03-02T05:15:17Z"
    message: ReplicaSet "kubeserve-dfbbf9445" has successfully progressed.
    reason: NewReplicaSetAvailable
    status: "True"
    type: Progressing
  observedGeneration: 1
  readyReplicas: 1
  replicas: 1
  updatedReplicas: 1
~
~
"/tmp/kubectl-edit-545655180.yaml" 66L, 1800C
```

Figure 4.12 – kubectl describing a Deployment for live-editing

Then, you can also delete Deployments if you don't need them anymore with the `kubectl delete` command:

```
kubectl delete deployment melon-serve
```

The following output shows that the Deployment has been deleted successfully:

```
deployment.apps "kubeserve" deleted
```

With the deletion of the Deployment, the objects defined in that Deployment are also deleted, as they share the same lifecycle. In our third example, the deployed `nginx` pods are deleted, as we delete the `kubeserve` Deployment.

Learning about Deployments allows you to manage your application in a more effective way, update it as an entity easier, and roll it back to its previous versions. In the next section, we'll have a look at rolling updates and rollbacks.

Performing rolling updates and rollbacks

Rolling updates provide a way to update a Deployment to a newer version more effectively and efficiently. This way, you can update Kubernetes objects such as replicas and pods gradually with nearly zero downtime. In a nutshell, you may consider either using the `kubectl set image` command or going straight to updating a YAML manifest file. In this section, we will introduce `kubectl set image`, as it is very effective and handy to use in your actual CKA exam.

Rolling updates with kubectl

From here, we'll go through the steps of rolling updates with `kubectl`:

1. You can spin up a new Deployment, `kubeserve`, using the following command:

    ```
    kubectl create deployment kubeserve --image=nginx:latest
    ```

2. You can use `kubectl` to update the container image as follows:

    ```
    kubectl set image deployment/kubeserve nginx=nginx:1.18.0
    --record
    ```

> **Important note**
> `--record` flag records information about the updates so that it can be rolled back later.
> You can either use `--record` flag or `--record=true` flag.

With the preceding command, you will see the following output:

```
deployment.apps/kubeserve image updated
```

3. You can use the `kubectl describe` command to double-check whether your container image has updated successfully by typing the following command:

`kubectl describe deploy kubeserve`

Your output should be similar to the following, in *Figure 4.14*, where you can see that the image is set to `nginx:1.18.0`:

```
[root@cloudmelonplayground:~# k describe deploy kubeserve
Name:                   kubeserve
Namespace:              default
CreationTimestamp:      Fri, 04 Mar 2022 05:06:32 +0000
Labels:                 app=kubeserve
Annotations:            deployment.kubernetes.io/revision: 2
                        kubernetes.io/change-cause: kubectl set image deployment/kubeserve nginx=nginx:1.18.0 --record=true
Selector:               app=kubeserve
Replicas:               1 desired | 1 updated | 1 total | 1 available | 0 unavailable
StrategyType:           RollingUpdate
MinReadySeconds:        0
RollingUpdateStrategy:  25% max unavailable, 25% max surge
Pod Template:
  Labels:  app=kubeserve
  Containers:
   nginx:
    Image:         nginx:1.18.0
    Port:          <none>
    Host Port:     <none>
    Environment:   <none>
    Mounts:        <none>
  Volumes:         <none>
Conditions:
  Type           Status  Reason
  ----           ------  ------
  Available      True    MinimumReplicasAvailable
  Progressing    True    NewReplicaSetAvailable
OldReplicaSets:  <none>
NewReplicaSet:   kubeserve-6d9c49fbd6 (1/1 replicas created)
Events:
  Type    Reason             Age    From                   Message
  ----    ------             ----   ----                   -------
  Normal  ScalingReplicaSet  2m31s  deployment-controller  Scaled up replica set kubeserve-dfbbf9445 to 1
  Normal  ScalingReplicaSet  15s    deployment-controller  Scaled up replica set kubeserve-6d9c49fbd6 to 1
  Normal  ScalingReplicaSet  8s     deployment-controller  Scaled down replica set kubeserve-dfbbf9445 to 0
```

Figure 4.13 – kubectl describing kubeserve after updating the image

The `kubectl describe deploy` command comes in very handy when we are trying to check key information such as the container image, ports, and deployment-related events. This is also the case in the actual CKA exam – make sure you master the shortcut of this command, `k describe deploy`, which will help you work more effectively in the exam.

Rollback

Rollback allows us to revert to a previous state and a Deployment makes this super easy to achieve:

1. You can use the following `kubectl rollout` command to quickly recover if you need to perform a rollback:

    kubectl rollout undo deployments kubeserve

    Your output should look as follows:

    deployment.apps/kubeserve rolled back

2. Now, if you use the `kubectl describe deploy kubeserve` command, you will see the following output indicating that the image has been rolled back:

```
Name:                   kubeserve
Namespace:              default
CreationTimestamp:      Fri, 04 Mar 2022 05:06:32 +0000
Labels:                 app=kubeserve
Annotations:            deployment.kubernetes.io/revision: 5
Selector:               app=kubeserve
Replicas:               1 desired | 1 updated | 1 total | 1 available | 0 unavailable
StrategyType:           RollingUpdate
MinReadySeconds:        0
RollingUpdateStrategy:  25% max unavailable, 25% max surge
Pod Template:
  Labels:  app=kubeserve
  Containers:
   nginx:
    Image:        nginx:latest
    Port:         <none>
    Host Port:    <none>
    Environment:  <none>
    Mounts:       <none>
  Volumes:        <none>
Conditions:
  Type           Status  Reason
  ----           ------  ------
  Available      True    MinimumReplicasAvailable
  Progressing    True    NewReplicaSetAvailable
OldReplicaSets:  <none>
NewReplicaSet:   kubeserve-dfbbf9445 (1/1 replicas created)
Events:
  Type    Reason             Age                 From                    Message
  ----    ------             ----                ----                    -------
  Normal  ScalingReplicaSet  2m14s               deployment-controller   Scaled up replica set kubeserve-6d9c49fbd6 to 1
  Normal  ScalingReplicaSet  75s                 deployment-controller   Scaled down replica set kubeserve-6d9c49fbd6 to 0
  Normal  ScalingReplicaSet  43s                 deployment-controller   Scaled up replica set kubeserve-5c5c66bc97 to 1
  Normal  ScalingReplicaSet  36s (x2 over 2m7s)  deployment-controller   Scaled down replica set kubeserve-dfbbf9445 to 0
  Normal  ScalingReplicaSet  6s (x3 over 4m30s)  deployment-controller   Scaled up replica set kubeserve-dfbbf9445 to 1
  Normal  ScalingReplicaSet  4s                  deployment-controller   Scaled down replica set kubeserve-5c5c66bc97 to 0
```

Figure 4.14 – kubectl describing kubeserve after a rollback

3. Now, you may be very curious as to whether we can keep a track of the history of our Deployments. You can use the following command:

    kubectl rollout history deployment kubeserve

The output would look as follows:

```
[root@cloudmelonplayground:~# kubectl rollout history deployment kubeserve
deployment.apps/kubeserve
REVISION   CHANGE-CAUSE
2          kubectl set image deployment/kubeserve nginx=nginx:1.18.0 --record=true
4          <none>
5          <none>
```

Figure 4.15 – kubectl describing kubeserve

4. In the case that you want to go back to a specific revision, you can use the --to-revision flag. You can see in *Figure 4.16* that we have revision 2 available thanks to using the --record flag when setting the image version. The following command is an example of undoing a Deployment and reverting to revision 2:

    kubectl rollout undo deployment kubeserve --to-revision=2

    Your output should look as follows:

    deployment.apps/kubeserve rolled back

5. Now, if you use the kubectl describe deploy kubeserve command, you will see the following output indicating that the image has been rolled back to revision 2:

```
[root@cloudmelonplayground:~# k describe deploy kubeserve
Name:                   kubeserve
Namespace:              default
CreationTimestamp:      Fri, 04 Mar 2022 05:06:32 +0000
Labels:                 app=kubeserve
Annotations:            deployment.kubernetes.io/revision: 6
                        kubernetes.io/change-cause: kubectl set image deployment/kubeserve nginx=nginx:1.18.0 --record=true
Selector:               app=kubeserve
Replicas:               1 desired | 1 updated | 1 total | 1 available | 0 unavailable
StrategyType:           RollingUpdate
MinReadySeconds:        0
RollingUpdateStrategy:  25% max unavailable, 25% max surge
Pod Template:
  Labels:  app=kubeserve
  Containers:
   nginx:
    Image:        nginx:1.18.0
    Port:         <none>
    Host Port:    <none>
    Environment:  <none>
    Mounts:       <none>
  Volumes:        <none>
Conditions:
  Type           Status  Reason
  ----           ------  ------
  Available      True    MinimumReplicasAvailable
  Progressing    True    NewReplicaSetAvailable
OldReplicaSets:  <none>
NewReplicaSet:   kubeserve-6d9c49fbd6 (1/1 replicas created)
Events:
  Type    Reason             Age                   From                    Message
  ----    ------             ----                  ----                    -------
  Normal  ScalingReplicaSet  2m21s                 deployment-controller   Scaled down replica set kubeserve-6d9c49fbd6 to 0
  Normal  ScalingReplicaSet  109s                  deployment-controller   Scaled up replica set kubeserve-5c5c66bc97 to 1
  Normal  ScalingReplicaSet  72s (x3 over 5m36s)   deployment-controller   Scaled up replica set kubeserve-dfbbf9445 to 1
  Normal  ScalingReplicaSet  70s                   deployment-controller   Scaled down replica set kubeserve-5c5c66bc97 to 0
  Normal  ScalingReplicaSet  7s (x2 over 3m20s)    deployment-controller   Scaled up replica set kubeserve-6d9c49fbd6 to 1
  Normal  ScalingReplicaSet  5s (x3 over 3m13s)    deployment-controller   Scaled down replica set kubeserve-dfbbf9445 to 0
```

Figure 4.16 – kubectl describing kubeserve

Deployments not only make the rolling update and rollback process much easier but also help us scale up and down with ease – we'll take a look at how to scale applications, as well as all the viable options when doing so, in the next section.

Scaling applications

When our application becomes popular, in order to handle increasingly on-demand requests, we need to spin up multiple instances of applications to satisfy the workload requirements.

When you have a Deployment, scaling is achieved by changing the number of replicas. Here, you can scale a Deployment using the `kubectl scale` command to make this happen:

```
kubectl scale deployment kubeserve --replicas=6
```

Your output should look as follows:

```
deployment.apps/kubeserve scaled
```

If you use the `kubectl get pods` command now, you will see that some more copies of the pods are spinning up, as shown in the following output:

```
[root@cloudmelonplayground:~# k get pods
NAME                        READY   STATUS      RESTARTS   AGE
hello-27441821-k9lzp        0/1     Completed   0          2m7s
hello-27441822-jhfp2        0/1     Completed   0          67s
hello-27441823-87n2p        0/1     Completed   0          7s
kubeserve-dfbbf9445-6vmkr   1/1     Running     0          61m
kubeserve-dfbbf9445-8hxxv   1/1     Running     0          61m
kubeserve-dfbbf9445-92m4c   1/1     Running     0          61m
kubeserve-dfbbf9445-cb9kz   1/1     Running     0          3m23s
kubeserve-dfbbf9445-tvmql   1/1     Running     0          61m
kubeserve-dfbbf9445-vf69g   1/1     Running     0          61m
```

Figure 4.17 – kubectl getting the pods and showing more copies of them

Aside from manually scaling the Deployments with the `kubectl scale` command, we also have another way of scaling a Deployment and its ReplicaSets, which is **HorizontalPodAutoscaler** (**HPA**). Let's take a look at the ReplicaSets first.

ReplicaSets

ReplicaSets help pods achieve higher availability since users can define a certain number of replicas using a ReplicaSet. The main capability of a ReplicaSet is to make sure the cluster keeps the exact number of replicas running in the Kubernetes cluster. If any of them were to fail, new ones would be deployed.

The following is an example of the YAML definition of a ReplicaSet:

```
apiVersion: apps/v1
kind: ReplicaSet
metadata:
  name: frontend
  labels:
    app: melonapp-rs
spec:
  replicas: 3
  selector:
    matchLabels:
      app: melonapp-rs
  template:
    metadata:
      labels:
        app: melonapp-rs
    spec:
      containers:
      - name: nginx
        image: nginx
```

The `matchLabels` selector simply matches the labels specified under it to the labels on the pods. To check your ReplicaSet, use the following command:

```
kubectl get replicaset
```

Alternatively, you can also use the following command:

```
kubectl get rs
```

Then, you will see the output indicating the number of DESIRED *replica* counts and how many of them are in a READY state:

```
[^Croot@cloudmelonplayground:~# k get rs
NAME                     DESIRED   CURRENT   READY   AGE
frontend                 3         3         3       46s
kubeserve-5c5c66bc97     0         0         0       42h
kubeserve-6d9c49fbd6     0         0         0       42h
kubeserve-cd44878f5      0         0         0       41h
kubeserve-dfbbf9445      6         6         6       42h
```

Figure 4.18 – The kubectl get rs command showing the state of the ReplicaSet

Once the ReplicaSet is deployed, update the number of ReplicaSets by using the following command:

```
kubectl scale replicaset frontend --replicas=6
```

Your output should look as follows:

```
replicaset.apps/frontend scaled
```

Alternatively, you can specify it in a YAML definition with the following command:

```
kubectl scale --replicas=6 -f replicas.yaml
```

Your output should look as follows:

```
replicaset.apps/frontend scaled
```

Now, if you want to check whether the number of ReplicaSets has increased, you can use the `kubectl get rs` command again and you will be able to see the following output:

Figure 4.19 – kubectl getting the ReplicaSets

In the case that you want to delete a ReplicaSet, you can use the `kubectl delete` command – in this case, we can use it to delete a ReplicaSet named `frontend`:

```
kubectl delete replicaset frontend
```

Your output should look as follows:

```
replicaset.apps "frontend" deleted
```

Using ReplicaSets directly is not the only way to scale the applications. Let's take a look at the alternative next, HPA.

HPA

To update a workload resource such as a Deployment or a StatefulSet, we can also use HPA – this is a Kubernetes API primitive that scales the workloads automatically based on your demands. *Figure 4.18* explains how HPA works in the context of application scaling:

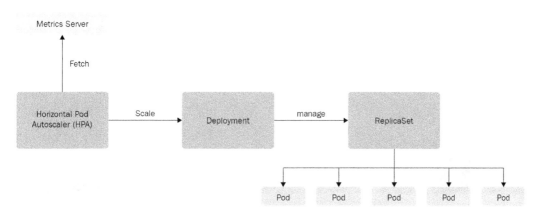

Figure 4.20 – HPA

From the previous diagram, we can see that HPA is configured to fetch metrics provided by a metrics server based on the CPU and memory usage. These metrics are fetched from `kubelet` by the metrics server, which then exposes them to the API server using a metrics API. HPA scales the Deployment by increasing or decreasing the count of replicas, which is managed underneath by a ReplicaSet.

As on-demand resource requests increase, HPA scales out the Deployment and the number of replicas increases. Conversely, when the resource requests decrease, the number of replicas decreases.

To create an HPA, you can use the `kubectl autoscale deployment` command with the following flags for the requirements:

- `cpu-percent` indicates the average CPU utilization usage across all pods
- `min` provides the minimum number of replicas
- `max` provides the maximum number of replicas

You can use the following command to create an HPA with a CPU utilization usage of 50% and ensure a minimum of 3 copies and a maximum of up to `10` copies:

```
kubectl autoscale deployment kubeserve --cpu-percent=50 --min=3
--max=10
```

Your output should look as follows:

```
horizontalpodautoscaler.autoscaling/kubeserve autoscaled
```

To check how many HPAs we currently have in the default namespace, use the following command:

```
kubectl get hpa
```

The output would look as follows

```
root@cloudmelonplayground:~# k get hpa
NAME        REFERENCE             TARGETS        MINPODS   MAXPODS   REPLICAS   AGE
kubeserve   Deployment/kubeserve  <unknown>/50%  3         10        0          10s
```

Figure 4.21 – Getting the HPAs in the default namespace

You can also use the following YAML definition to deploy an HPA, which will help you achieve the same goal:

```
apiVersion: autoscaling/v2
kind: HorizontalPodAutoscaler
metadata:
  name: kubeserve
spec:
  scaleTargetRef:
    apiVersion: apps/v1
    kind: Deployment
    name: kubeserve
  minReplicas: 3
  maxReplicas: 10
  metrics:
  - type: Resource
    resource:
      name: cpu
      target:
        type: Utilization
        averageUtilization: 50
```

In the case that you want to delete an HPA, use a `kubectl delete` command. Here, we can delete an HPA named `kubeserve` as follows:

```
kubectl delete hpa kubeserve
```

Your output will look as follows:

```
horizontalpodautoscaler.autoscaling "kubeserve" deleted
```

Another concept that we will cover is DaemonSets, which come in handier in real life, particularly in scenarios where at least one replica of the pod needs to be evenly distributed across the worker nodes. Let's get right into it.

DaemonSets

We have learned about how ReplicaSets and Deployments help us ensure that multiple copies of our applications are up and running across various worker nodes. DaemonSets create a couple of copies of a pod, meanwhile making sure that at least one copy of the pod is evenly on each node in the Kubernetes cluster, as shown in *Figure 4.23*.

If a new node is added to the cluster, a replica of that pod is automatically assigned to that node. Similarly, when a node is removed, the pod is automatically removed.

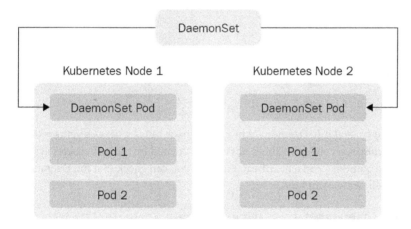

Figure 4.22 – DaemonSets

You can define a DaemonSet using the following YAML definition:

```
apiVersion: apps/v1
kind: DaemonSet
metadata:
  name: fluentd
  namespace: kube-system
  labels:
    k8s-app: fluentd
spec:
  selector:
    matchLabels:
      name: fluentd
  template:
    metadata:
      labels:
        name: fluentd
    spec:
      containers:
      - name: fluentd
        image: fluentd:latest
```

Your output would look as follows:

```
daemonset.apps/fluentd created
```

Notice that we have created this DaemonSet in a namespace called kube-system this time – this is a namespace usually reserved for Kubernetes objects created by the Kubernetes system. We'll get to talking about the namespace in a heartbeat. For now, you can check that the DaemonSet has been created using the following command:

```
kubectl get daemonsets -n kube-system
```

Alternatively, we can simplify the command:

```
kubectl get ds -n kube-system
```

Your output will look as follows:

```
root@cloudmelonplayground:~# k get ds -n kube-system
NAME         DESIRED    CURRENT    READY    UP-TO-DATE    AVAILABLE    NODE SELECTOR             AGE
fluentd      1          1          1        1             1            <none>                    28s
kube-proxy   1          1          1        1             1            kubernetes.io/os=linux    9d
```

Figure 4.23 – Checking out the DaemonSets in the kube-system namespace

Don't forget to check the details of the DaemonSets by using the following:

```
kubectl describe daemonsets fluentd -n kube-system
```

Your output would look as follows:

```
[root@cloudmelonplayground:~# k describe ds fluentd -n kube-system
Name:            fluentd
Selector:        name=fluentd
Node-Selector:   <none>
Labels:          k8s-app=fluentd
Annotations:     deprecated.daemonset.template.generation: 1
Desired Number of Nodes Scheduled: 1
Current Number of Nodes Scheduled: 1
Number of Nodes Scheduled with Up-to-date Pods: 1
Number of Nodes Scheduled with Available Pods: 1
Number of Nodes Misscheduled: 0
Pods Status:  1 Running / 0 Waiting / 0 Succeeded / 0 Failed
Pod Template:
  Labels:   name=fluentd
  Containers:
   fluentd:
    Image:          fluentd:latest
    Port:           <none>
    Host Port:      <none>
    Environment:    <none>
    Mounts:         <none>
  Volumes:          <none>
Events:
  Type      Reason            Age     From                    Message
  ----      ------            ---     ----                    -------
  Normal    SuccessfulCreate  56s     daemonset-controller    Created pod: fluentd-rrl4m
```

Figure 4.24 – kubectl describing the DaemonSets

In case you want to delete a DaemonSet, use the kubectl delete command. Here, we can delete a DaemonSet named fluentd in the kube-system namespace as follows:

```
kubectl delete ds fluentd -n kube-system
```

Your output should look as follows:

```
daemonset.apps "fluentd" deleted
```

The main use case of DaemonSets is to use them as a monitoring agent or a logs collector on every node, or in other cases, to run a cluster storage daemon across all the worker nodes.

With DaemonSets, you don't have to worry about removing or adding new nodes that will impact the monitoring agents on these nodes. A real-life use case, such as `fluentd`, requires an agent to be deployed on each node in the cluster.

Workload scheduling

Understanding the workload scheduling and how it works with the Kubernetes scheduler will be useful in your daily life as a Kubernetes Administrator. Kubernetes allows you to define node affinity rules, taints, and tolerations with the good use of labels, selectors, and annotations leading your way. Let's first start with the notion of namespaces.

Understanding namespaces

Thinking about the separation of the workloads, namespaces come in handy. A namespace is a logical separation of all the namespaced objects deployed in a single Kubernetes cluster. Deployments, Services, and Secrets are all namespaced. Otherwise, some Kubernetes objects are cluster-wide, such as Nodes, StorageClass, and PersistentVolume. The name of a resource has to be unique within a namespace.

You can get all namespaces using the following command:

```
kubectl get namespaces
```

Alternatively, you can use this command:

```
kubectl get ns
```

You will see that the output gets all the namespace currently in our Kubernetes cluster:

```
root@cloudmelonplayground:~# k get ns
NAME              STATUS    AGE
default           Active    10d
kube-node-lease   Active    10d
kube-public       Active    10d
kube-system       Active    10d
```

Figure 4.25 – kubectl getting the namespaces

When you define a pod or any namespaced Kubernetes object, you can specify the namespace in the YAML definition as follows:

```
apiVersion: v1
kind: Pod
metadata:
  name: k8s-ns-pod
  namespace: k8s-ns
  labels:
    app: k8sapp
spec:
  containers:
  - name: k8sapp-container
    image: busybox
    command: ['sh', '-c', 'echo Salut K8S! && sleep 3600']
```

If you create that pod and specify the namespace that the pod belongs to, you can add the -n flag when querying this pod using the kubectl get pods command. The following is an example:

```
kubectl get pods -n k8s-ns
```

Similarly, if the pod has been created in that namespace, you can use the following command to check it out:

```
kubectl describe pod k8s-ms-pod -n k8s-ns
```

In the case that the pods are not in the default namespace, you don't have to specify the namespace option anymore. In the following example, you want to set a namespace named dev, and then use the kubectl get command without the -n flag:

```
kubectl config set-context &(kubectl config current-context)
--namespace=dev
```

You can then simply run the following command without the namespace option to list the pods:

```
kubectl get pods
```

Understanding namespaces will further help you when you need to define the namespace-scoped permissions where Kubernetes objects are grouped. We'll elaborate on this further in *Chapter 6, Securing Kubernetes*.

Labels, node selectors, and annotations

Labels, selectors, and annotations are useful notions when it comes to workload scheduling. Labels are key-value pairs attached to Kubernetes objects that can be listed in the `metadata.labels` section of an object descriptor. Selectors are used for identifying and selecting a group of objects using their labels. See the following examples of some quality-based selectors:

```
kubectl get pods -l app=my-app
kubectl get pods -l environment=production
```

When it comes to inequality, you can use the following:

```
kubectl get pods -l environment!=production
```

The following example involves chaining multiple selectors together using a comma-delimited list:

```
kubectl get pods -l app=myapp.environment=production
```

To assign a pod to nodes, we can use node selectors. You can specify a map of key-value pairs in the `PodSpec` field:

You can start by labeling the worker nodes using the following command:

```
kubectl label node cloudmelonplayground env=dev
```

The output should be as follows:

```
node/cloudmelonplayground labeled
```

You can use the following command to show the label of worker nodes:

```
kubectl get nodes --show-labels
```

Then, we should get the following output:

```
NAME                   STATUS    ROLES               AGE   VERSION    LABELS
cloudmelonplayground   Ready     control-plane,master   10d   v1.23.3    beta.kubernetes.io/arch=arm64,beta.kubernetes.io/os=linux,env=dev,kuber
netes.io/arch=arm64,kubernetes.io/hostname=cloudmelonplayground,kubernetes.io/os=linux,minikube.k8s.io/commit=362d5fdc0a3dbee389b3d3f1034e8023
e72bd3a7,minikube.k8s.io/name=minikube,minikube.k8s.io/primary=true,minikube.k8s.io/updated_at=2022_02_25T01_55_58_0700,minikube.k8s.io/versio
n=v1.25.2,node-role.kubernetes.io/control-plane=,node-role.kubernetes.io/master=,node.kubernetes.io/exclude-from-external-load-balancers=
```

Figure 4.26 – Getting the node labels

Then, you can add the node selector in the YAML definition as follows:

```
apiVersion: v1
kind: Pod
metadata:
```

```
    name: nginx
    labels:
       env: test
  spec:
    containers:
    - name: nginx
       image: nginx
    nodeSelector:
       env: dev
```

We can attach annotations to objects using the `metadata.annotations` section, as with the following configuration file that has the annotation `imageregistry: "http://hub.docker.com/"`:

```
apiVersion: v1
kind: Pod
metadata:
  name: melon-annotation
  annotations:
     imageregistry: "https://hub.docker.com/"
spec:
  containers:
  - name: nginx
     image: nginx:latest
     ports:
     - containerPort: 80
```

Annotations are similar to labels and they can be used to store custom metadata about objects.

Node affinity and anti-affinity

Node affinity and anti-affinity are simply ways to help pods be assigned to the right node. Compare this to `nodeSelector`, which is designed for assigning a pod directly to the worker nodes. The following is an example of node affinity and anti-affinity in the YAML specification:

```
  spec:
   affinity:
     podAffinity:
```

```
        requiredDuringSchedulingIgnoredDuringExecution:
        - labelSelector:
            matchExpressions:
            - key: security
              operator: In
              values:
              - S1
          topologyKey: topology.kubernetes.io/zone
      podAntiAffinity:
        preferredDuringSchedulingIgnoredDuringExecution:
        - weight: 100
          podAffinityTerm:
            labelSelector:
              matchExpressions:
              - key: security
                operator: In
                values:
                - S2
            topologyKey: topology.kubernetes.io/zone
```

With particular labels, node affinity and anti-affinity allow us to create matching rules with logic and operations.

Taints and tolerations

Aside from node affinity and anti-affinity, we can also assign taints on the node and tolerations on the pods by tainting the nodes and ensuring that no pods will be scheduled to that node.

You can use the following command to taint a node:

```
kubectl taint nodes melonnode app=melonapp:NoSchedule
```

The preceding definition can be translated into a pod YAML definition file to achieve the same outcome as follows:

```
apiVersion: v1
kind: Pod
metadata:
 name: melon-ns-pod
 namespace: melon-ns
```

```
    labels:
      app: melonapp
  spec:
   containers:
   - name: melonapp-container
     image: busybox
     command: ['sh', '-c', 'echo Salut K8S! && sleep 3600']
   tolerations:
   - key: "app"
     operator: "Equal"
     value: "melonapp"
     effect: "NoSchedule"
```

If you want to un-taint a node, you can use the following command:

```
kubectl taint nodes yourworkernode    node-role.kubernetes.io/
yourworkernode:NoSchedule-
```

We have learned how to taint certain nodes when you want to evict workloads from a node in this section. Now, let's look at resource management.

Resource management

Kubernetes allows us to specify the resource requirements of a container in the pod specification, which basically refers to how many resources a container needs.

`kube-scheduler` uses the resource request information that you specify for a container in a pod to decide on which worker node to schedule the pod. It's up to `kubelet` to enforce these resource limits when you specify them for the containers in the pod so that the running container goes beyond a set limit, as well as reserves at least the requested amount of a system resource for a container to use.

It usually gives us the following values:

- `resources.limits.cpu` is the resource limit set on CPU usage.

- `resources.limits.memory` is the resource limit set on memory usage.

- `resources.requests.cpu` is the minimum CPU usage requested to allow your application to be up and running.

- `resources.requests.memory` is the minimum memory usage requested to allow your application to be up and running. In the case that a container exceeds its memory request, the worker node that it runs on becomes short on overall memory at the same time, and the pod that the container belongs to is likely to be evicted too.

- `resources.limits.ephemeral-storage` is the limit on ephemeral storage resources.

- `resources.limits.hugepages-<size>` is the limit on the allocation and consumption of pre-allocated huge pages by any applications in a pod.

A resource request refers to the amount of resources that are necessary to run a container, and what they do is govern on which worker node the containers will actually be scheduled. So, when Kubernetes is getting ready to run a particular pod, it's going to choose a worker node based on the resource requests of that pod's containers. Kubernetes will use these values to ensure that it chooses a node that actually has enough resources available to run that pod. A pod will only run on a node that has enough available resources to run the pod's containers. The following is a YAML example of defining resource request and limits:

```
apiVersion: v1
kind: Pod
metadata:
 name: melonapp-pod
spec:
 containers:
 - name: melonapp-container
   image: busybox
   command: ['sh', '-c', 'echo stay tuned! && sleep 3600']
   resources:
     requests:
       memory: "64Mi"    # 64 Megabytes
       cpu: "250m"
     limits:
       memory: "128Mi"
       cpu: "500m"
```

You can use the `kubectl describe node` command to check the allocation resources of that node to see whether your requests or limits definitions correspond to what is needed in the current circumstances:

```
Allocated resources:
  (Total limits may be over 100 percent, i.e., overcommitted.)
  Resource              Requests     Limits
  _____              _____     _____

  cpu                   850m (42%)   0 (0%)
  memory                370Mi (9%)   170Mi (4%)
  ephemeral-storage     0 (0%)       0 (0%)
  hugepages-1Gi         0 (0%)       0 (0%)
  hugepages-2Mi         0 (0%)       0 (0%)
  hugepages-32Mi        0 (0%)       0 (0%)
  hugepages-64Ki        0 (0%)       0 (0%)
Events:                 <none>
```

Figure 4.27 – kubectl describing the node resources

You can use the `kubectl top` command in the case that you have a metrics server installed in your cluster to check the actual resource usage of the node or pod.

Configuring applications

Configuring an application is a simple and straightforward experience thanks to ConfigMaps and Secrets. Let's take a look at each of them.

Understanding ConfigMaps

A ConfigMap is simply a Kubernetes object that stores configuration data in key-value pairs. This configuration data can then be used to configure the software running in a container by configuring a pod to consume ConfigMaps using environment variables, command-line arguments, or mounting a volume with configuration files.

You can also use a YAML definition to define `configmap` as follows:

```
apiVersion: v1
kind: ConfigMap
metadata:
  name: melon-configmap
data:
  myKey: myValue
  myFav: myHome
```

Your output should look as follows:

```
configmap/melon-configmap created
```

You can check `configmap` using the following command:

```
kubectl get configmap
```

Alternatively, you can use this command:

```
kubectl get cm
```

Your output will be as follows:

```
[root@cloudmelonplayground:~# k get configmap
NAME               DATA   AGE
kube-root-ca.crt   1      10d
melon-configmap    2      77s
```

Figure 4.28 – kubectl getting configmap

You can check the binary data of `configmap` using the following command:

```
k describe configmap melon-configmap
```

The following screenshot is the output of the preceding command:

```
Name:          melon-configmap
Namespace:     default
Labels:        <none>
Annotations:   <none>

Data
====
myFav:
----
myHome
myKey:
----
myValue

BinaryData
====

Events:  <none>
```

Figure 4.29 – The configmap binary data

Once you have `configmap` ready, here's how to configure the pod to consume it:

1. Create a pod that can use the `configmap` data by using environment variables:

```
apiVersion: v1
kind: Pod
metadata:
  name: melon-configmap
spec:
  containers:
  - name: melonapp-container
image: busybox
    command: ['sh', '-c', "echo $(MY_VAR) && sleep 3600"]
    env:
    - name: MY_VAR
      valueFrom:
        configMapKeyRef:
          name: melon-configmap
          key: myKey
```

You can use the following command to check the `configmap` value:

```
kubectl logs melon-configmap
```

The output will be similar to the following:

```
root@cloudmelonplayground:~# kubectl logs melon-configmap
myValue
```

Figure 4.30 – The configmap mounted value

2. You can create a pod to use `configmap` data via a volume. The following is an example of a YAML definition:

```
apiVersion: v1
kind: Pod
metadata:
  name: melon-volume-pod
spec:
  containers:
    - name: myapp-container
```

```
        image: busybox
        command: ['sh', '-c', "echo $(cat /etc/config/myKey)
&& sleep 3600"]
        volumeMounts:
          - name: config-volume
            mountPath: /etc/config
    volumes:
        - name: config-volume
          configMap:
            name: melon-configmap
```

You can use the `kubectl logs` command to check the pod for the mounted data value, or use the following command to check the `configmap`:

```
kubectl exec melon-volume-pod -- ls /etc/config
```

The output will look as follows:

```
^Croot@cloudmelonplayground:~# kubectl exec melon-volume-pod -- ls /etc/config
myFav
myKey
```

Figure 4.31 – The configmap mounted value

In the case that you want to delete a `configmap`, use the `kubectl delete` command:

```
kubectl delete cm melon-configmap
```

Your output would look as follows:

```
configmap "melon-configmap" deleted
```

Here, we have shown how we can work with ConfigMaps in Kubernetes. Once you feel comfortable with ConfigMaps, you'll find a lot of similarities when it comes to working with Secrets. Next, we will have a look at how to work with Kubernetes Secrets so that they can be consumed by your application.

Understanding Secrets

A Kubernetes Secret is an object containing sensitive data such as a password, an API token, or a key, which is passed to a pod rather than stored in a `PodSpec` field or in the container itself:

```
kubectl create melon-secret --from-literal=username=packtuser
  --from-literal=password='S!B\*d$zDsb='
```

You can also use a YAML definition to define `configmap` as the following with base64:

```yaml
apiVersion: v1
kind: Secret
metadata:
  name: melon-secret
type: Opaque
data:
  USER_NAME: bXllc2VybmFtZTQo=
  PASSWORD: bXlwYXNzd29yZAo=
```

You can check the Secrets by using the following command:

```
kubectl get secrets
```

Your output would look as follows:

```
root@cloudmelonplayground:~# k get secrets
NAME                  TYPE                                  DATA   AGE
default-token-sp8z4   kubernetes.io/service-account-token   3      10d
melon-secret          Opaque                                2      11s
```

Figure 4.32 – kubectl getting Secrets

Once you have created the Secret, you may want to attach the Secret to an application. That's where you need to create a pod to consume the Secret by following these steps:

1. You can create a pod to consume the Secret using environment variables:

    ```yaml
    apiVersion: v1
    kind: Pod
    metadata:
      name: melon-secret-pod
    spec:
      containers:
        - name: test-container
          image: busybox:latest
          command: [ "/bin/sh", "-c", "env" ]
          envFrom:
          - secretRef:
              name: melon-secret
      restartPolicy: Never
    ```

2. You can also consume a Secret as a volume, as shown here – you will define a `secret-volume` and then mount the `secret-volume` to the `/etc/secret-volume` path:

```
volumes:
- name: secret-volume
  secret:
    secretName: melon-secret
containers:
- name: mybusybox
  image: busybox:latest
  command: [ "/bin/sh", "-c", "env" ]
  volumeMounts:
  - name: secret-volume
    readOnly: true
    mountPath: "/etc/secret-volume"
```

If you want to delete a Secret, use the `kubectl delete` command as follows:

```
kubectl delete secret melon-secret
```

Your output will look as follows:

```
secret "melon-secret" deleted
```

Note that if you delete a Secret, make sure to update the `PodSpec` field for your application to avoid exceptions. You can do this by creating a new Secret, then attaching it to your pod, or updating your application so it doesn't need the Secret anymore.

Manifest management with kustomize

Starting from Kubernetes 1.14, customization files became available to facilitate smoother Kubernetes management. It supports the following use cases:

-   Generation of YAML definitions from other resources, such as generating a Kubernetes Secret and its YAML definition

-   Common configuration across multiple YAML definitions, such as adding namespace for a group of resources

-   Composing and customizing a collection of YAML definitions, such as setting resource requests and limits for multiple Kubernetes objects

This can be achieved using a central file called Kustomization.yaml. You can use the following command to view the resources found in the directory that are contained in a customization file:

```
kubectl kustomize <targeting_kustomization_directory>
```

You can then apply those resources by running the following command:

```
kubectl apply -k <targeting_kustomization_directory>
```

Let's take Secret generation as an example and generate a Secret manifest file:

```
# Create a password.txt file
cat <<EOF >./password.txt
username=admin
password=secret
EOF
cat <<EOF >deployment.yaml
apiVersion: apps/v1
kind: Deployment
metadata:
  name: my-app
  labels:
    app: my-app
spec:
  selector:
    matchLabels:
      app: my-app
  template:
    metadata:
      labels:
        app: my-app
    spec:
      containers:
      - name: app
        image: my-app
        volumeMounts:
        - name: password
          mountPath: /secrets
      volumes:
```

```
        - name: password
          secret:
            secretName: example-secret-1
EOF
cat <<EOF >./kustomization.yaml
resources:
- deployment.yaml
secretGenerator:
- name: example-secret-1
  files:
  - password.txt
EOF
```

Then, you will be able to see that you have two files created after executing the previous steps:

```
kustomization.yaml   password.txt
```

If you want to check out the content of the `customization.yaml` file, you can use `cat customization.yaml` and you will see the following output:

```
secretGenerator:
- name: example-secret-1
  files:
  - password.txt
```

Then, you can use the `kubectl apply` command to deploy the pod with the Secret mounted:

```
kubectl apply -f ./test
```

Kustomize is a good way to customize your application configuration and now that it is built into `kubectl apply -k`, you can gain a greater understanding of the use cases of Kustomize by visiting the official documentation site: `https://kubectl.docs.kubernetes.io/guides/`.

Common package management and templating with Helm

Helm is a management tool for managing packages of pre-configured Kubernetes objects in the form of charts – we call these Helm charts. Helm charts allow users to install and manage Kubernetes applications more reproducibly and effectively. Furthermore, you can find popular Helm charts from the community or share your own applications with the Helm community at this link: `https://artifacthub.io/packages/search`.

The standard file structure of a chart is as follows:

- `Charts` – (the folder)

- `Chart.yaml #` – A .yaml file that contains the information about the chart

- `README.md`

- `requirements.lock`

- `requirements.yaml` – an optional file that lists the dependencies for a chart (the dependencies are actually packaged in the `Charts` folder)

- `templates` – a directory of templates that combine with values to generate Kubernetes manifest files

- `values.yaml` – contains the default configuration values for the chart (this is where Helm grabs the values for the manifest template that contains the reference values)

To query the Helm charts that have been deployed, use the following command:

```
helm install stable/melonchart
```

If you need to search for a chart, you can use the following command:

```
helm search chartname
```

Delete a Helm chart that has been deployed using the following command:

```
helm delete melonchart
```

Whenever you install a chart, a new release is created. So, one chart can be installed multiple times into the same cluster. Each can be independently managed and upgraded. To upgrade a release to a specified version of a chart or update the chart values, run the following:

```
helm upgrade [RELEASE] [CHART_path] [flags]
```

To roll back to a specific version, you can use the following command:

```
helm rollback melon-release 2
```

Helm charts help you manage, install, and upgrade Kubernetes-native applications. You can learn more about Helm by visiting their official documentation website: `https://helm.sh/docs/`.

Summary

In this chapter, we covered one of the most common tasks for both Kubernetes Administrators and Developers – application scheduling and managing the application lifecycle. Even though this chapter covers about 15% of the content of the CKA exam, working with Kubernetes objects is one of the most important daily tasks as a Kubernetes Administrator. Ensure that you practice enough and master the shortcuts of the kubectl commands before moving on.

In the next chapter, we'll talk about Kubernetes storage. The content and the questions covered in *Chapter 4, Application Scheduling and Lifecycle Management* and *Chapter 5, Demystifying Kubernetes Storage* are considered very high-value and less time-consuming within the scheme of the actual CKA exam. Stay tuned and keep learning!

Mock CKA scenario-based practice test

You have two virtual machines, master-0 and worker-0. Please complete the following mock scenarios.

Scenario 1

SSH into the worker-0 node and provision a new pod called ngnix with a single container, nginx.

Scenario 2

SSH to worker-0 and then scale nginx to 5 copies.

Scenario 3

SSH to worker-0, set a ConfigMap with a username and password, and then attach a new pod to BusyBox.

Scenario 4

SSH to worker-0 and create a nginx pod with an init container called busybox.

Scenario 5

SSH to worker-0, create a nginx pod, and then a busybox container in the same pod.

You can find all the scenario resolutions in *Appendix - Mock CKA scenario-based practice test resolutions* of this book.

FAQs

- *Where can I find out about Helm charts?*

  Go to Helm's official documentation to learn more about Helm: `https://helm.sh/docs/howto/charts_tips_and_tricks/`.

- *Where can I find out about Kustomize?*

  Go to Helm's official documentation to learn more about Kustomize: `https://kubectl.docs.kubernetes.io/references/kustomize/`.

- *What is the recommended official Kubernetes article about init containers?*

  I recommend bookmarking this article, *Init Containers*: `https://kubernetes.io/docs/concepts/workloads/pods/init-containers/`.

- *What is your recommended Kubernetes official article for ConfigMaps?*

  I recommend bookmarking an article, *ConfigMaps*: `https://kubernetes.io/docs/concepts/configuration/configmap/`.

- *What is your recommended official Kubernetes article for resource management?*

  I recommend bookmarking this article, *Resource Management for Pods and Containers*: `https://kubernetes.io/docs/concepts/configuration/manage-resources-containers/`.

5

Demystifying Kubernetes Storage

In this chapter, we will discuss the core concept of Kubernetes storage for stateful workloads and shows how to configure applications with mounted storage and dynamically persistent storage. This chapter covers 10% of the **Certified Kubernetes Administrator (CKA)** exam content.

In this chapter, we're going to cover the following main topics:

- Stateful versus stateless workloads

- Kubernetes volumes

- Kubernetes StorageClasses

- Volume modes, access modes, and reclaim policies for volumes

- Configuring an application with mounted storage

- Configuring an application with persistent storage

Technical requirements

To get started, we need to make sure your local machine meets the following technical requirements:

- A compatible Linux host – we recommend a Debian-based Linux distribution such as Ubuntu 18.04 or later

- Make sure your host machine has at least 2 GB RAM, 2 CPU cores, and about 20 GB of free disk space

Stateful versus stateless workloads

Kubernetes is designed for both stateful and stateless applications. To maintain stateless workloads in Kubernetes, we can freely delete and replace containers without any additional concerns. The stateful application usually has storage attached either locally or in a remote location, as it needs to hold client data. That data could be short-lived or *non-persistent* storage, which means that it is just maintained until the expiration of a session. An example of this is the Redis cache on Kubernetes. Another use case is when the data needs to be held for long enough by using persistent storage so that it can be used on-demand. An example of the latter is the MongoDB operator for Kubernetes. The whole story is much more complicated than it seems but it all starts with Kubernetes volumes.

Kubernetes volumes represent the concept of storage in Kubernetes. As mentioned in *Chapter 1, Kubernetes Overview*, the volumes in Kubernetes are managed by storage drivers tailored by storage vendors. This part is no longer part of Kubernetes source code after the **Container Storage Interface (CSI)** was introduced.

A volume can support local storage, on-premises software-defined storage, cloud-based storage (such as blob, block, or file storage), or a **network file system** (**NFS**) as shown in *Figure 5.1*:

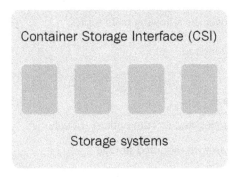

Figure 5.1 – A CSI

Then, users can use CSI-compatible volume drivers and CSI volumes to attach or directly mount the pods up and running in the Kubernetes cluster.

Kubernetes volumes

Ephemeral volumes and persistent volumes are two main types of volumes in Kubernetes. We'll take a look at each of them. Some of them may not be covered in the CKA exam, but it is important to know, as whichever organization you work in will have embarked on its journey with one of those public cloud providers.

Ephemeral storage

Ephemeral volumes targeted to the application need to hold the data, but they don't care about data loss in the case that the pod fails or restarts – the lifecycle of the ephemeral volume is aligned with the pod lifecycle. With that in mind, mounted storage is usually ephemeral, as it shares the same lifecycle as your containers. As long as the container is stopped or destroyed during the process of restarting the pod, any internal storage is completely removed.

Another use case is when a pod contains multiple containers. It is possible to mount that storage to the containers and allow those containers to share the same volume so that they interact with the same shared filesystem.

Ephemeral volumes have several types, which we will cover one by one.

emptyDir

emptyDir is one of the most common types of ephemeral storage and will appear in the CKA exam. It usually serves as an empty directory when the pod starts, and it shares the same lifecycle with the Pod, meaning it only exists as long as a pod is up and running, and the data in the emptyDir is deleted permanently when the pod stops or restarts.

When it comes to multi-containers in the same pod, it can be shared across containers, although each container can mount the emptyDir in a different repository, as shown in *Figure 5.2*:

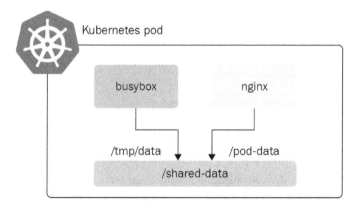

Figure 5.2 – Multi-containers in a pod sharing storage volumes

The following is an example YAML definition of an emptyDir mounted to a pod:

```
apiVersion: v1
kind: Pod
metadata:
  name: multi-containers
spec:
  restartPolicy: Never
  volumes:
  - name: shared-data
    emptyDir: {}
  containers:
  - name: busybox-pod
    image: busybox
    command: ["/bin/sh","-c","while true; do sleep 3600; done"]
    volumeMounts:
    - name: shared-data
      mountPath: /tmp/data
  - name: nginx-pod
    image: nginx
    volumeMounts:
    - name: shared-data
      mountPath: /data
```

Through the preceding example, you can see how to mount shared volumes between two containers, which would come in handy when you want those two containers to consume the same data source.

CSI ephemeral volumes

CSI ephemeral volumes are CSI driver-compatible volumes that serve as temporary storage. For a very long time in the past, CSI volumes provided by an external storage driver in Kubernetes were used as persistent volumes, with the goal of not sharing a lifecycle with the pod. Starting from Kubernetes 1.15, CSI drivers can also be used for such ephemeral inline volumes. The following is an example of using CSI ephemeral volumes:

```
kind: Pod
apiVersion: v1
metadata:
  name: my-csi-pod
```

```
spec:
  containers:
    - name: my-frontend
      image: busybox
      volumeMounts:
      - mountPath: "/data"
        name: my-csi-vol
      command: ["sleep", "1000000"]
  volumes:
    - name: my-csi-vol
      csi:
        driver: inline.storage.kubernetes.io
        volumeAttributes:
          foo: bar
```

These CSI storage drivers are generally third-party, such as Azure Disk, Azure File, AWS EBS, and DellEMC unity – you can find a complete list of CSI drivers at `https://kubernetes-csi.github.io/docs/drivers.html`.

Generic ephemeral volumes

Generic ephemeral volumes are general drivers with some additional features available such as snapshotting, storage cloning, storage resizing, and storage capacity tracking. The following is an example of using CSI ephemeral volumes:

```
kind: Pod
apiVersion: v1
metadata:
  name: my-app
spec:
  containers:
    - name: my-frontend
      image: busybox
      volumeMounts:
      - mountPath: "/cache"
        name: cache-volume
      command: [ "sleep", "1000000" ]
  volumes:
    - name: scratch-volume
```

```
  ephemeral:
    volumeClaimTemplate:
      metadata:
        labels:
          type: my-cache-volume
      spec:
        accessModes: [ "ReadWriteOnce" ]
        storageClassName: "my-cache-storage-class"
        resources:
          requests:
            storage: 1Gi
```

Generic ephemeral volumes work with all storage drivers that support dynamic provisioning, including some third-party CSI storage drivers. Now that we have a good understanding of ephemeral volumes, we'll have a look at projected volumes and see how they work with Kubernetes.

Projected volumes

Configuration data is mounted to the Kubernetes pods – this data was injected into a pod through the sidecar pattern. We covered `ConfigMap` and `Secret` objects in *Chapter 4, Application Scheduling and Lifecycle Management,* which fall under this category. More specifically, they are also called **projected volumes** because they represent a volume that maps several existing volumes into the same directory.

Besides ConfigMap and Secret, projected volumes also consist of **downwardAPI** volumes and **service account tokens**. We'll take a closer look at them here with some examples.

A `downwardAPI` volume is designed to make downward API data available to applications. Similarly, it also mounts as a directory and then writes the data in plain-text files. The downward API allows containers to consume cluster or pod information without using the Kubernetes API server or through the client.

The following example shows you how to mount `downwardAPI` as a projected volume:

```
apiVersion: v1
kind: Pod
metadata:
  name: volume-test
spec:
  containers:
  - name: container-test
    image: busybox
```

```
    volumeMounts:
    - name: all-in-one
      mountPath: "/projected-volume"
      readOnly: true
  volumes:
  - name: all-in-one
    projected:
      sources:
      - downwardAPI:
          items:
            - path: "labels"
              fieldRef:
                fieldPath: metadata.labels
            - path: "cpu_limit"
              resourceFieldRef:
                containerName: container-test
                resource: limits.cpu
```

A service account token type of projected volume is designed to make downward API data available to applications. Similarly, it also mounts as a directory and then writes the data in plain-text files.

The following example shows you how to mount a service account token as a projected volume:

```
apiVersion: v1
kind: Pod
metadata:
  name: volume-test
spec:
  containers:
  - name: container-test
    image: busybox
    volumeMounts:
    - name: all-in-one
      mountPath: "/projected-volume"
      readOnly: true
  volumes:
  - name: all-in-one
    projected:
```

```
sources:
- serviceAccountToken:
    audience: api
    expirationSeconds: 3600
    path: token
```

Let's wrap up what we covered in this section about downwardAPI and service account token volumes, as well as recall what we learned about ConfigMap and Secret objects in *Chapter 4, Application Scheduling and Lifecycle Management*, by looking at the following. This is an all-in-one example to help you understand how to work with all of them in one encounter:

```
apiVersion: v1
kind: Pod
metadata:
  name: volume-test
spec:
  containers:
  - name: container-test
    image: busybox
    volumeMounts:
    - name: all-in-one
      mountPath: "/projected-volume"
      readOnly: true
  volumes:
  - name: all-in-one
    projected:
      sources:
      - secret:
          name: mysecret
          items:
            - key: username
              path: my-group/my-username
      - downwardAPI:
          items:
            - path: "labels"
              fieldRef:
                fieldPath: metadata.labels
```

```
            - path: "cpu_limit"
              resourceFieldRef:
                containerName: container-test
                resource: limits.cpu
      - configMap:
          name: myconfigmap
          items:
            - key: config
              path: my-group/my-config
      - serviceAccountToken:
          audience: api
          expirationSeconds: 3600
          path: token
```

All the projected volumes, `configMap`, `downwardAPI`, `secret`, plus `emptyDir`, are provided as local ephemeral storage. On each node, `kubelet` is in charge of provisioning and managing pods, and managing the local ephemeral storage.

Aside from the mounted storage serving as internal storage, in some use cases, we also need persistent data outside the life of the container itself that continues to exist even if the container stops or is replaced. This raises the requirement to have permanent external storage assigned to our pods. We'll take a look at persistent volumes in the next section.

Persistent storage

Compared to ephemeral volumes, persistent volumes have a lifecycle that is independent of the Kubernetes pods. State persistence means keeping some data or information to continue beyond the life of the container when the container is deleted or replaced. However, it can be modified or updated by the containers while it's running.

The mechanism of working with persistent volume in Kubernetes takes advantage of the exposed API, which abstracts technical details of how the external storage is provided, as well as how it is consumed. Kubernetes allows us to work with persistent storage through the notion of persistent volumes and persistent volume claims:

- A **PersistentVolume** (**PV**) is a storage resources provisioned dynamically based on the storage classes with a set of features to fulfill the user's requirements.

- A **PersistentVolumeClaim** (**PVC**) is the abstraction layer between the pod and the PV requested by the user, with a set of requirements including the specific level of resources and the access modes.

As shown in the following, *Figure 5.3*, the PV and PVC are defined in the Kubernetes cluster, while the physical storage is outside of the Kubernetes cluster:

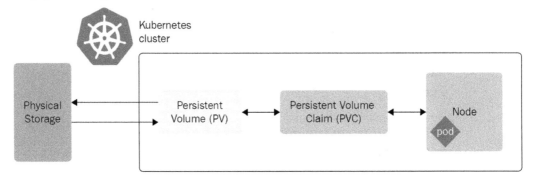

Figure 5.3 – The PV and PVC

Equally, note that the PV can be bound to a PVC, and it is a cluster-wide resource, while the PVC is namespaced.

Let's cover some other important concepts with regards to working with a PV and PVC before we dive into *how*.

The StorageClass

The `StorageClass` resource in Kubernetes classifies the Kubernetes storage class. As a matter of fact, a `StorageClass` contains a `provisioner`, `parameters`, and `reclaimPolicy` field.

The provisioner represents which CSI volume plugin is being used to provision the PVs. Examples of different provisioners are Azure Disk, AWS EBS, and Glusterfs. You can find a complete list of supported `StorageClass` resources here: `https://kubernetes.io/docs/concepts/storage/storage-classes/`.

We need to define the storage class in the PVC and the definition of storage classes includes the provisioner and the reclaim policy. Their relationship is shown in *Figure 5.3*:

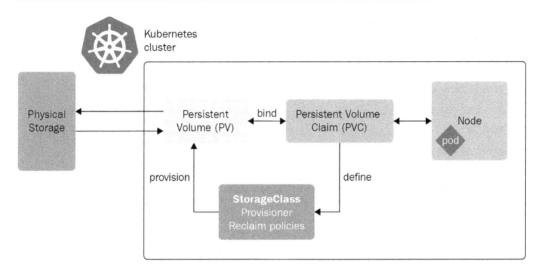

Figure 5.4 – A StorageClass resource

Notice that when the reclaim policy is not specified, it defaults to `Delete`, which means if a user deletes the PVC that is bound to this PV, the PVC itself gets deleted too. You can also set it to `Retain`, which means it will be retained and that you will need to manually delete the data that resides in it. Another case would be to set it to `Recycle`. In this case, the PV will be recycled, deprecated, and replaced by dynamic provisioning, which will depend on the provisioner. The DefaultStorageClass admission controller on the Kubernetes API server will also need to be enabled – this is out of the scope of the CKA exam but I think it's still worth a mention here.

The following is an example `StorageClass` definition, using an Azure Disk-managed disk to define a `StorageClass` resource with a YAML definition:

```
apiVersion: storage.k8s.io/v1
kind: StorageClass
metadata:
  name: slow
provisioner: kubernetes.io/azure-disk
parameters:
  storageaccounttype: Standard_LRS
  kind: managed
```

Interestingly, despite the fact that local volumes don't support dynamic provisioning, they can still be created and bound when the pod is scheduled. We can set `volumeBindingMode` to `WaitForFirstConsumer`, which is shown as follows:

```
apiVersion: storage.k8s.io/v1
kind: StorageClass
metadata:
  name: local-storage
commissions: kubernetes.io/no commissions
volumeBindingMode: WaitForFirstConsumer
```

Learning about storage class in Kubernetes will help you work with different storage in real life, going above and beyond what's required in the current CKA exam. Please feel free to check out the official documentation – it will be updated whenever a new supported storage class is added and will provide useful examples: `https://kubernetes.io/docs/concepts/storage/storage-classes/`.

Now, let's take a look at another important concept called volume modes next.

Volume modes

Volume modes indicate the type of consumption of the volume – this can either be a filesystem or a block device. When `volumeMode` is set to `Filesystem`, it mounts into the pods as a directory. When `volumeMode` is set to `Block`, we use it as a raw block.

Access modes

When a PV is mounted to a pod, we can specify different access modes. The access modes represent the way that the data in the storage resources is being consumed. They can be summarized as shown in the following table:

Access modes	Definition	Abbreviated
ReadWriteOnce	The volume can be mounted as read-write by one node.	RWO
ReadOnlyMany	The volume can be mounted as read only by multiple nodes.	ROX
ReadWriteMany	The volume can be mounted as read-write by multiple nodes.	RWX
ReadWriteOncePod	The volume can be mounted as read-write by one pod. This is a feature supported by Kubernetes, starting from Kubernetes 1.22.	RWOP

To learn more about access modes, you can find the official documentation here: `https://` `kubernetes.io/docs/concepts/storage/persistent-volumes/#access-modes`.

Knowing the access modes is important, as they're used all the time when working with Kubernetes storage. Now, let's take a look at the PV and PVC next, and see how these concepts work together with Kubernetes.

A PV

Let's first take a look at how to create a PV. You do so using the following YAML definition:

```
apiVersion: v1
kind: PersistentVolume
metadata:
  name: my-pv
spec:
  storageClassName: local-storage
  capacity:
    storage: 1Gi
  accessModes:
    - ReadWriteOnce
```

To learn more about how PVs work with Kubernetes, check out this article: `https://kubernetes.` `io/docs/concepts/storage/persistent-volumes/#persistent-volumes`.

Knowing about PVs on their own is not enough – we need to learn about how PVCs work alongside them within Kubernetes storage, which is what we'll get into next.

PVCs

One of the most interesting things about the PVC is that users don't need to worry about the details of where the storage is located. They only need to know about the `StorageClass` and `accessMode`. PVCs will automatically bind themselves to a PV that has a compatible `StorageClass` and `accessMode`. The following is an example of a PVC:

```
apiVersion: v1
kind: PersistentVolumeClaim
metadata:
  name: my-pvc
spec:
  storageClassName: local-storage
```

```
accessModes:
   - ReadWriteOnce
resources:
  requests:
    storage: 512Mi
```

You can learn more about the PVC from the official Kubernetes documentation here: `https://kubernetes.io/docs/concepts/storage/persistent-volumes/#lifecycle-of-a-volume-and-claim`.

Once you have a PV and PVC that will define the Kubernetes storage, the next step is to assign the storage to your applications deployed on top of Kubernetes. As we explained, Kubernetes is also capable of dealing with stateful workloads, so we'll have a look at how to mount storage to a stateful application in Kubernetes.

Cracking stateful applications in Kubernetes

In this section, we will learn about how to work with storage for stateful applications in Kubernetes. The considerations within this part are often seen as high-value and low-effort in terms of the CKA exam. Make sure you keep practicing them until you feel you know them confidently:

- Mounting storage to a stateful application
- Dynamically provisioning storage to a stateful application

Configuring an application with mounted storage

You need to create a new YAML definition where you write up the specification of the Kubernetes pod and then set up emptyDir volumes for the pod. Kubernetes creates empty storage on a node after the pod is scheduled to a specific worker node:

1. Check whether you currently have any nodes available to schedule a pod by using the following command:

   kubectl get nodes

   Alternatively, you can use the simplified version of the previous command:

   alias k=kubectl

   k get no

   If the status of any of your nodes shows Ready, as in the following figure, that means you can proceed to the next step:

```
[root@cloudmelonplayground:~# k get no
NAME                     STATUS    ROLES                    AGE    VERSION
cloudmelonplayground     Ready     control-plane,master     16h    v1.23.3
```

Figure 5.5 – Checking the available nodes

2. Use the Vim editor to create a new YAML definition file called `pod-volume.yaml`, and when you enter Vim, press the *Insert* key on your keyboard and let the current `edit` mode switch to `INSERT`:

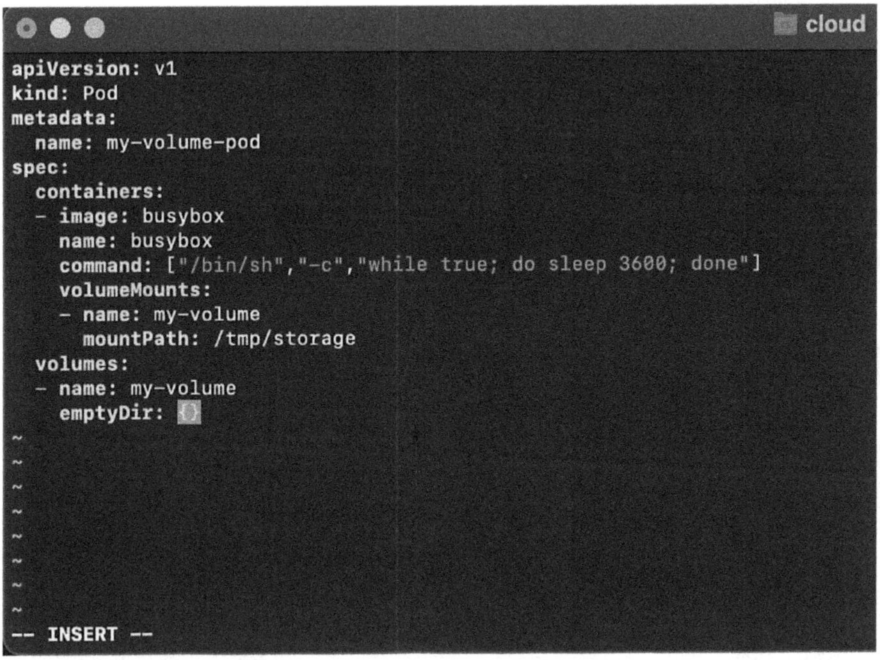

```
apiVersion: v1
kind: Pod
metadata:
  name: my-volume-pod
spec:
  containers:
  - image: busybox
    name: busybox
    command: ["/bin/sh","-c","while true; do sleep 3600; done"]
    volumeMounts:
    - name: my-volume
      mountPath: /tmp/storage
  volumes:
  - name: my-volume
    emptyDir: {}
~
~
~
~
~
~
~
-- INSERT --
```

Figure 5.6 – Inserting a YAML spec with Vim

3. Then, put the following in the YAML definition:

```
apiVersion: v1
kind: Pod
metadata:
  name: my-volume-pod
spec:
  containers:
  - image: busybox
    name: busybox
```

```
    command: ["/bin/sh","-c","while true; do sleep 3600;
done"]
    volumeMounts:
    - name: my-volume
      mountPath: /tmp/storage
  volumes:
  - name: my-volume
    emptyDir: {}
```

4. Then, save your edits and quit Vim. Press the *Esc* key, type :wq! at the bottom of the editor, and then press *Enter* to take you back to the terminal:

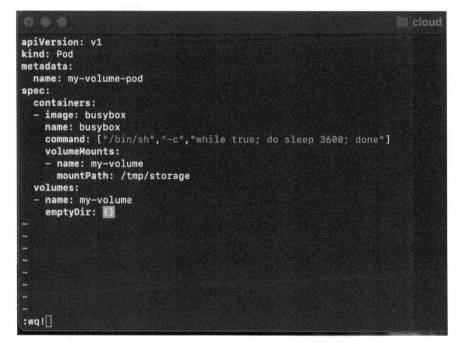

Figure 5.7 – Saving the YAML definition in Vim

5. When you're on the terminal, use the following command to deploy the .yaml file:

```
kubectl apply -f pod-volume.yaml
```

Then, it should display a message that the pod has been created successfully, something similar to the following:

```
pod/my-volume-pod created
```

You can go ahead and check whether the pod is now running by using the `kubectl get pods` command and the command comes back with the following output:

```
root@cloudmelonplayground:~# kubectl get pods
NAME              READY   STATUS    RESTARTS   AGE
my-volume-pod     1/1     Running   0          30m
```

Figure 5.8 – Checking whether the pod is running

Now, you have deployed a pod with mounted storage. If you run the following command, you'll be able to check out further details, including configuration information, resource requirements, the labels of the pods, and events information about this pod and the mounted storage:

```
kubectl describe pod my-volume-pod
```

The output of this command should be similar to the following:

```
root@cloudmelonplayground:~# k describe pod my-volume-pod
Name:         my-volume-pod
Namespace:    default
Priority:     0
Node:         cloudmelonplayground/172.16.16.132
Start Time:   Fri, 25 Feb 2022 17:56:06 +0000
Labels:       <none>
Annotations:  <none>
Status:       Running
IP:           172.17.0.3
IPs:
  IP:  172.17.0.3
Containers:
  busybox:
    Container ID:   docker://bbc4ab0af39f481dbea0e9adbd4d53f575386d61e4ff87e89329211b2b18b670
    Image:          busybox
    Image ID:       docker-pullable://busybox@sha256:afcc7f1ac1b49db317a7196c902e61c6c3c4607d63599ee1a82d702d249a0ccb
    Port:           <none>
    Host Port:      <none>
    Command:
      /bin/sh
      -c
      while true; do sleep 3600; done
    State:          Running
      Started:      Fri, 25 Feb 2022 17:56:11 +0000
    Ready:          True
    Restart Count:  0
    Environment:    <none>
    Mounts:
      /tmp/storage from my-volume (rw)
      /var/run/secrets/kubernetes.io/serviceaccount from kube-api-access-8p6p6 (ro)
Conditions:
  Type              Status
  Initialized       True
  Ready             True
  ContainersReady   True
  PodScheduled      True
Volumes:
  my-volume:
    Type:       EmptyDir (a temporary directory that shares a pod's lifetime)
    Medium:
    SizeLimit:  <unset>
  kube-api-access-8p6p6:
    Type:                    Projected (a volume that contains injected data from multiple sources)
    TokenExpirationSeconds:  3607
    ConfigMapName:           kube-root-ca.crt
    ConfigMapOptional:       <nil>
    DownwardAPI:             true
QoS Class:                   BestEffort
Node-Selectors:              <none>
Tolerations:                 node.kubernetes.io/not-ready:NoExecute op=Exists for 300s
                             node.kubernetes.io/unreachable:NoExecute op=Exists for 300s
Events:
  Type    Reason     Age   From               Message
  ----    ------     ----  ----               -------
  Normal  Scheduled  25s   default-scheduler  Successfully assigned default/my-volume-pod to cloudmelonplayground
  Normal  Pulling    24s   kubelet            Pulling image "busybox"
  Normal  Pulled     20s   kubelet            Successfully pulled image "busybox" in 3.989513851s
  Normal  Created    20s   kubelet            Created container busybox
  Normal  Started    20s   kubelet            Started container busybox
```

Figure 5.9 – Checking the pod configurations and status

From the output, we can see the pod has been mounted on a volume called `my-volume` just as we specified in the YAML definition. `Type` has been specified as `EmptyDir`, so it's a temporary directory that shares the pod's lifecycle. The bottom of the screenshot also shows the relevant events when provisioning this pod.

Configuring an application with persistent storage

In this case, you need to create a new YAML definition where you write up the specification of the Kubernetes PV – Kubernetes will assign the storage based on the PVC bound to the PV on a node after the pod has been scheduled to a specific worker node.

Creating your PV

You can start by checking whether you currently have any nodes available to schedule a pod by using `kubectl get nodes` or `kubectl get no`. Make sure that the status of one of your nodes is `Ready`, as in the following:

```
[root@cloudmelonplayground:~# k get no
NAME                   STATUS    ROLES                    AGE    VERSION
cloudmelonplayground   Ready     control-plane,master     16h    v1.23.3
```

Figure 5.10 – Checking the available nodes

From here, we're creating a new PV by going through the following steps:

1. Use Vim to write up the following YAML definition called `data-pv.yaml`:

    ```yaml
    apiVersion: v1
    kind: PersistentVolume
    metadata:
      name: data-pv
    spec:
      storageClassName: local-storage
      capacity:
        storage: 1Gi
      accessModes:
        - ReadWriteOnce
      hostPath:
        path: "/mnt/data"
    ```

2. When you're on the terminal, use the following command to deploy the .yaml file:

    kubectl apply -f data-pv.yaml

    Then, it will display a message that the PV has been created successfully, something similar to the following:

    ```
    persistentvolume/data-pv created
    ```

    The preceding YAML definition means that there is 1 GB of storage allocated as local storage. You can define a PVC of 1 G storage bound to that PV. However, in the theoretical case that you had two claims of 500 MB each, the PV could also be split during the allocation process. Under the hood, those two PVCs are bound to the same PV, and from there, they share the amount of storage.

3. Use the following command to check the PV's status:

    kubectl get pv

    You'll get the following output:

```
root@cloudmelonplayground:~# k get pv
NAME      CAPACITY   ACCESS MODES   RECLAIM POLICY   STATUS      CLAIM   STORAGECLASS    REASON   AGE
data-pv   1Gi        RWO            Retain           Available           local-storage            3s
```

Figure 5.11 – Checking whether the PV is available

Notice that the status is `available`, meaning that this PV is currently not bound to a PVC and is available to be bound with a new PVC, which we're about to create in the next step.

Creating your PVC

From here, we're creating a new PVC by going through the following steps:

1. Use Vim to write up the following YAML definition called `data-pvc.yaml`:

    ```
    apiVersion: v1
    kind: PersistentVolumeClaim
    metadata:
      name: data-pvc
    spec:
      storageClassName: local-storage
      accessModes:
        - ReadWriteOnce
      resources:
        requests:
          storage: 512Mi
    ```

2. When you're on the terminal, use the following command to deploy the yaml file:

    `kubectl apply -f data-pvc.yaml`

    The PVC is created successfully and gives an output similar to the following:

    `persistentvolumeclaim/data-pvc created`

3. Use the following command to check the PVC's status:

    `kubectl get pvc`

You'll get the following output:

```
root@cloudmelonplayground:~# k get pvc
NAME        STATUS   VOLUME    CAPACITY   ACCESS MODES   STORAGECLASS    AGE
data-pvc    Bound    data-pv   1Gi        RWO            local-storage   4s
```

Figure 5.12 – Checking the PVC

You may notice that the status of this PVC is Bound, which means that it is bound to a PV.

To double-check whether it is bound to the PV that you desire, you can use `kubectl get pv` command to check back:

```
root@cloudmelonplayground:~# k get pv
NAME       CAPACITY   ACCESS MODES   RECLAIM POLICY   STATUS   CLAIM              STORAGECLASS    REASON   AGE
data-pv    1Gi        RWO            Retain           Bound    default/data-pvc   local-storage            13m
```

Figure 5.13 – Check whether the PVC is bound to the PV

The preceding figure shows the Bound status of our PV, which means it has been bound successfully.

Configuring the pod to consume the PV

From here, we're configuring the pod to consume the PV by going through the following steps:

1. Use Vim to write up the following YAML definition called `data-pod.yaml` where we're about to create a pod to consume the targeted PV:

    ```
    apiVersion: v1
    kind: Pod
    metadata:
      name: data-pod
    spec:
      containers:
        - name: busybox
    ```

```
        image: busybox
        command: ["/bin/sh", "-c","while true; do sleep
3600;  done"]
        volumeMounts:
        - name: temp-data
          mountPath: /tmp/data
    volumes:
        - name: temp-data
          persistentVolumeClaim:
            claimName: data-pvc
    restartPolicy: Always
```

2. 2. When you're on the terminal, use the following command to deploy the yaml file:

kubectl apply -f data-pod.yaml

The pod is successfully created with an output similar to the following:

```
pod/data-pod created
```

You can use the kubectl get pods command to verify whether your pod is up and running. If you want your command to watch the status of the pod, you can use the -w flag in your command; it should look as follows:

kubectl get pods -w

The output would look as follows:

```
[root@cloudmelonplayground:~# k get pods
NAME        READY    STATUS                 RESTARTS    AGE
data-pod    0/1      ContainerCreating      0           4s
[root@cloudmelonplayground:~# k get pods -w
NAME        READY    STATUS     RESTARTS    AGE
data-pod    1/1      Running    0           10s
```

Figure 5.14 – Checking whether the pod is up and running

You can use the following command to check out further details, including configuration information, resource requirements, labels of the pods, and event information about this pod and the dynamically allocated storage:

kubectl describe pod data-pod

The output of this command should be similar to the following:

```
[root@cloudmelonplayground:~# k describe pod data-pod
Name:           data-pod
Namespace:      default
Priority:       0
Node:           cloudmelonplayground/192.168.239.128
Start Time:     Fri, 25 Feb 2022 23:41:42 +0000
Labels:         <none>
Annotations:    <none>
Status:         Running
IP:             172.17.0.3
IPs:
  IP:   172.17.0.3
Containers:
  busybox:
    Container ID:   docker://6b82670cb125b6864b1934565cbae81678eab2f027b708083634daeefe00752b
    Image:          busybox
    Image ID:       docker-pullable://busybox@sha256:afcc7f1ac1b49db317a7196c902e61c6c3c4607d63599ee1a82d702d249a0ccb
    Port:           <none>
    Host Port:      <none>
    Command:
      /bin/sh
      -c
      while true; do sleep 3600; done
    State:          Running
      Started:      Fri, 25 Feb 2022 23:41:46 +0000
    Ready:          True
    Restart Count:  0
    Environment:    <none>
    Mounts:
      /tmp/data from temp-data (rw)
      /var/run/secrets/kubernetes.io/serviceaccount from kube-api-access-159z4 (ro)
Conditions:
  Type              Status
  Initialized       True
  Ready             True
  ContainersReady   True
  PodScheduled      True
Volumes:
  temp-data:
    Type:       PersistentVolumeClaim (a reference to a PersistentVolumeClaim in the same namespace)
    ClaimName:  data-pvc
    ReadOnly:   false
  kube-api-access-159z4:
    Type:                    Projected (a volume that contains injected data from multiple sources)
    TokenExpirationSeconds:  3607
    ConfigMapName:           kube-root-ca.crt
    ConfigMapOptional:       <nil>
    DownwardAPI:             true
QoS Class:        BestEffort
Node-Selectors:   <none>
Tolerations:      node.kubernetes.io/not-ready:NoExecute op=Exists for 300s
                  node.kubernetes.io/unreachable:NoExecute op=Exists for 300s
Events:
  Type    Reason     Age   From               Message
  ----    ------     ----  ----               -------
  Normal  Scheduled  25s   default-scheduler  Successfully assigned default/data-pod to cloudmelonplayground
  Normal  Pulling    24s   kubelet            Pulling image "busybox"
  Normal  Pulled     21s   kubelet            Successfully pulled image "busybox" in 3.162937658s
  Normal  Created    21s   kubelet            Created container busybox
  Normal  Started    21s   kubelet            Started container busybox
```

Figure 5.15 – Checking the pod's detailed configuration and events

From the output, we can see the pod has been dynamically attached to persistent storage called temp-data, which was expected, as we defined it in the YAML definition. The bottom of the screenshot also shows the relevant events while provisioning this pod.

The preceding is an example of using a PVC as a volume – this allows pods to access storage by using the claim as a volume. In that case, the claim must exist in the same namespace in which the pods will be using them.

We also noticed that, in some cases, people use host Path to mount volumes, which simply allocates local storage of that node of the cluster so that the pod consumes the storage where the pod lives.

> **Important note**
>
> host Path also easily causes security issues, so we should avoid using it as much as possible. While using it, we can specify volumeMounts as ReadOnly and only make it available to a specific file or repository.

The following is an example of this:

```
apiVersion: v1
kind: Pod
metadata:
  name: my-pv
  namespace: default
spec:
  restartPolicy: Never
  volumes:
  - name: vol
    hostPath:
      path: /tmp/data
  containers:
  - name: my-pv-hostpath
    image: "busybox"
    command: ["/bin/sh", "-c","while true; do sleep 3600; done"]
    volumeMounts:
    - name: vol
      mountPath: /scrub
```

Note that host Path works for a single node only, and if you're on a multi-node cluster, a local volume is the way to go. You can find more details about local storage at https://kubernetes.io/docs/concepts/storage/volumes/#local.

Summary

This chapter covers one of the highest-value topics in the CKA exam, which is Kubernetes storage. Over the last three years, the CKA exam has raised more and more attention toward Kubernetes storage, where it previously only scratched the surface and now focuses on various use cases of the stateful application deployment. Learning this part may not seem the most crucial for Kubernetes administrators at the moment, but it will take off more quickly once we have more and more cloud-native databases adopted by enterprise-grade customers. Having a solid knowledge of storage will add value to your existing Kubernetes administration skills. If you can confidently play with the exercises in this chapter, it will increase your success rate in the actual CKA exam, as storage-related questions are usually simpler but higher value compared to other cluster maintenance task-related questions in the previous chapters.

In the next chapter, *Securing Kubernetes*, we will dive into some important Kubernetes security concepts, which will help you not only set up a solid foundation for the CKA exam but also potentially help you for the **Certified Kubernetes Security Specialist (CKS)** exam in the future – stay tuned!

Mock CKA scenario-based practice test

You have two virtual machines, `master-0` and `worker-0`. Please complete the following mock scenarios.

Scenario 1

Create a new PV called `packt-data-pv` to store 2 GB, and two PVCs each requesting 1 GB of local storage.

Scenario 2

Provision a new pod called `pack-storage-pod` and assign an available PV to this Pod.

You can find all the scenario resolutions in *Appendix - Mock CKA scenario-based practice test resolutions* of this book.

FAQs

- *Where can I find the latest updates about the supported CSI drivers while working with Kubernetes?*

  The Kubernetes CSI **Special Interest Group** (**SIG**) has a GitHub-based documentation website where you can find all the latest drivers, with tutorials from their main page: `https://kubernetes-csi.github.io/docs`. More specifically, you can find all available supported CSI drivers at the following link: `https://kubernetes-csi.github.io/docs/drivers.html`.

- *What is the recommended official Kubernetes article to refer to for configuring ephemeral storage?*

  I recommend bookmarking the official documentation about ephemeral volumes: `https://kubernetes.io/docs/concepts/storage/ephemeral-volumes/`.

- *What is the recommended official Kubernetes article to refer to for configuring persistent storage?*

  I recommend bookmarking this article, *Configure a Pod to Use a Persistent Volume for Storage*, where you can find all the key steps and processes: `https://kubernetes.io/docs/tasks/configure-pod-container/configure-persistent-volume-storage/`.

6
Securing Kubernetes

This chapter will cover how Kubernetes authentication and authorization patterns work and dive into Kubernetes **role-based access control** (**RBAC**). We'll also learn about managing the security of applications deployed on Kubernetes.

Since most of the Kubernetes security-related content released prior to November 2020 has gradually moved to the **Certified Kubernetes Security Specialist** (**CKS**) exam instead, this chapter will just cover the essentials to help you to learn about Kubernetes security. We'll specifically focus on Kubernetes RBAC since it is close to 5% of the CKA exam content.

Nonetheless, a good understanding of the Kubernetes security fundamentals will be a great help for the CKA exam and prepare you for further development in the Kubernetes space.

In this chapter, we're going to cover the following main topics:

- Securing Kubernetes in layers
- Kubernetes authentication and authorization
- Kubernetes RBAC
- Managing the security of Kubernetes applications

Technical requirements

To get started, you will need to make sure that your local machine meets the technical requirements described as follows:

- A compatible Linux host – we recommend a Debian-based Linux distribution such as Ubuntu 18.04 or later.
- Make sure that your host machine has at least 2 GB of RAM, 2 CPU cores, and about 20 GB of free disk space.

Securing Kubernetes in layers

Kubernetes security is a broad topic due to the sophistication of the platform. It includes secure Kubernetes nodes, networks, and Kubernetes objects such as Pods. The **Cloud Native Computing Foundation** (**CNCF**) defines Kubernetes security in layers, which they call the *four Cs* of cloud-native security, taking the topic of security beyond Kubernetes and its ecosystem. The four Cs stand for **Cloud**, **Cluster**, **Container**, and **Code**, as shown in the following diagram:

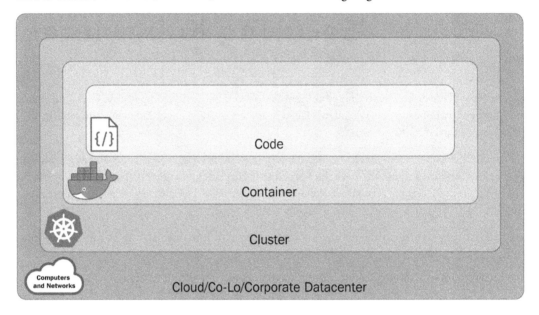

Figure 6.1 – The different layers in Kubernetes

From the preceding diagram, we can see the following:

- The **Cloud** layer is based on the underlying infrastructure where the Kubernetes cluster is deployed – it is managed by the cloud provider when it is in the cloud or by the organization when it comes to a private data center.

- The **Cluster** layer is more about securing the Kubernetes cluster components, ensuring each component is secured and conjured correctly. Looking back at *Chapter 1, Kubernetes Overview*, will help you understand how those components work together.

- The **Container** layer includes container vulnerability scanning, hosted OS scaling, and container privileged users.

- The **Code** layer is focused on the application code. Different from traditional application security approaches, it now works with DevSecOps and vulnerability assessment tools. This layer is relevant but outside of the scope of Kubernetes security.

Cloud-native security or, more specifically, Kubernetes security requires organizations to address each layer. In this chapter, we'll focus on the following topics:

- Kubernetes API security with an admission controller

- Kubernetes authentication and authorization with RBAC, **Attribute-Based Access Control** (**ABAC**), and node authorization

- Managing the security of Kubernetes applications with security contexts

The preceding topics are either part of cluster-layer or container-layer security, and they help us run our Kubernetes application securely. We'll cover Kubernetes network security and dive deeper into network policies in *Chapter 7, Demystifying Kubernetes Networking*.

Kubernetes authentication and authorization

In *Chapter 1, Kubernetes Overview*, we talked about a typical workflow of Kubernetes components collaborating with each other. In this workflow, when a request comes through the Kubernetes API server, it invokes an API call. This API request now needs to be authenticated and authorized by the API server before a request is made to a Kubernetes API resource. As a result, the request can either be *allowed* or *denied*. The authentication process can be depicted as in *Figure 6.2*:

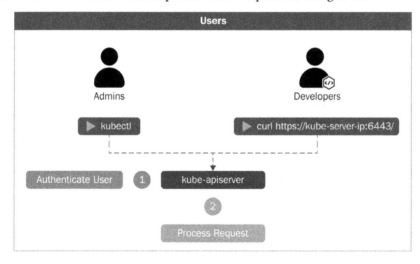

Figure 6.2 – API Kubernetes authentication

You can refer to the following article to get an overview of how the Kubernetes authentication process works: `https://kubernetes.io/docs/reference/access-authn-authz/authentication/`.

Before getting into authentication and authorization, let's take a look at the user accounts and service accounts in Kubernetes.

Service accounts versus user accounts

In Kubernetes, we have a distinction between normal user accounts and service accounts managed by Kubernetes. An account represents an identity for a user or a service process. The main difference between a user account and a service account is as follows:

- **User accounts** are for normal human users. In Kubernetes, the RBAC subsystem is used to determine whether the user is authorized to perform a specific operation on a specific scope. We'll look into this further in the *Kubernetes RBAC* section later in this chapter.

- **Service accounts** are for services or processes running in a Pod in the Kubernetes cluster. The service accounts are users managed by the Kubernetes API. In Kubernetes, it is possible to use client certificates, bearer tokens, or even an authenticating proxy to authenticate API requests through an API server.

We'll take a closer look at the following things from hereon:

- Kubernetes service accounts and how to work with them

- How to organize cluster access using `kubeconfig` as a Kubernetes user

- How to configure access to multiple clusters as a Kubernetes user

Let's take a look at the Kubernetes service account first.

Kubernetes service accounts

Back in the previous chapter, we created a new Pod with `kubectl`, although there is a default service account in the `default` namespace, which the Pod was actually automatically assigned to. Now, let's have a look at how to work with a service account in Kubernetes.

Managing service accounts

You can use the following command to get the current service account in the `default` namespace:

```
kubectl get serviceaccounts
```

Alternatively, you can simply use the shortcut command as follows:

```
kubectl get sa
```

The output will return the default service account in the `default` namespace:

```
NAME        SECRETS    AGE
default     1          5d
```

The service account is a namespaced resource – you can use the following command to check out all the service accounts in the current cluster:

```
k get sa -A
```

Alternatively, you can use the complete command as follows:

```
k get serviceaccounts --all-namespaces
```

The output of the preceding commands will list the service account information by namespace, similar to the following in *Figure 6.3*:

```
cloudmelon@cloudmelonplayground:~$ k get sa -A
NAMESPACE         NAME                                  SECRETS   AGE
default           default                               1         59d
fission-builder   default                               1         38d
fission-builder   fission-builder                       1         38d
fission-function  default                               1         38d
fission-function  fission-fetcher                       1         38d
fission-ns        default                               1         38d
fission           default                               1         38d
fission           fission-svc                           1         38d
kube-node-lease   default                               1         59d
kube-public       default                               1         59d
kube-system       attachdetach-controller               1         59d
kube-system       bootstrap-signer                      1         59d
kube-system       certificate-controller                1         59d
kube-system       clusterrole-aggregation-controller    1         59d
kube-system       coredns                               1         59d
kube-system       cronjob-controller                    1         59d
kube-system       daemon-set-controller                 1         59d
kube-system       default                               1         59d
kube-system       deployment-controller                 1         59d
kube-system       disruption-controller                 1         59d
kube-system       endpoint-controller                   1         59d
kube-system       endpointslice-controller              1         59d
kube-system       endpointslicemirroring-controller     1         59d
kube-system       ephemeral-volume-controller           1         59d
kube-system       expand-controller                     1         59d
kube-system       generic-garbage-collector             1         59d
kube-system       horizontal-pod-autoscaler             1         59d
kube-system       job-controller                        1         59d
kube-system       kube-proxy                            1         59d
kube-system       metrics-server                        1         3d16h
kube-system       namespace-controller                  1         59d
kube-system       node-controller                       1         59d
kube-system       persistent-volume-binder              1         59d
kube-system       pod-garbage-collector                 1         59d
kube-system       pv-protection-controller              1         59d
kube-system       pvc-protection-controller             1         59d
kube-system       replicaset-controller                 1         59d
kube-system       replication-controller                1         59d
kube-system       resourcequota-controller              1         59d
kube-system       root-ca-cert-publisher                1         59d
kube-system       service-account-controller            1         59d
kube-system       service-controller                    1         59d
kube-system       statefulset-controller                1         59d
kube-system       storage-provisioner                   1         59d
kube-system       token-cleaner                         1         59d
kube-system       ttl-after-finished-controller         1         59d
kube-system       ttl-controller                        1         59d
```

Figure 6.3 – The service account information by namespace

This also means we can get the service account information by namespace using the `kubectl get sa` command and then by specifying the -n flag with `namespace name` to get the service account with that particular namespace. For example, using `kubectl get sa -n` with `kube-system` will only return the service account in the `kube-system` namespace.

The kubectl create sa command

You can use the `kubectl create` command to create a new service account, the following being an example:

```
kubectl create serviceaccount melon-serviceaccount
```

The following output will show that the service account is created successfully:

```
serviceaccount/melon-serviceaccount created
```

We can also create the service account in a different namespace using the `kubectl create` command by specifying the -n flag. Additionally, we also need to make sure that the namespace exists prior to creating a service account in that namespace. The following is an example of using a `kubectl create` command to create a service account named `melonsa` in a namespace called `melon-ns`:

```
kubectl create ns melon-ns
kubectl create sa melonsa -n melon-ns
```

The preceding output displays that you have created the service account successfully. You can also use the following command to check that the service account has just been created:

```
k get -n melon-ns serviceaccounts
```

The following output lists the service account and how long it's been created:

```
NAME                 SECRETS    AGE
melon-ssa    1           46s
```

Similarly, if you want to check out the service account in another namespace, you can use the `kubectl get sa <service account name>` command and then add the -n flag, for example, `k get sa melonsa -n melon-ns`.

Assigning a service account to a Pod

The purpose of having a service account is to provide an identity to serve the process running in the Pod. To determine the service account that a Pod will use, you can specify a `serviceAccountName` field in the Pod YAML specification called `sa-pod.yaml`, as shown here:

```
apiVersion: v1
kind: Pod
metadata:
    name: melon-serviceaccount-pod
spec:
    serviceAccountName: melon-serviceaccount
    containers:
    - name: melonapp-svcaccount-container
      image: busybox
      command: ['sh', '-c','echo stay tuned!&& sleep 3600']
```

Then, when we use the `kubectl apply -f sa-pod.yaml` command to deploy this YAML file, we'll be able to see a Pod spinning up.

The kubectl delete sa command

You can delete a service account using the `kubectl delete sa <account name >` command:

```
kubectl delete sa melon-serviceaccount
```

The output comes back showing that the service account was deleted:

```
serviceaccount "melon-serviceaccount" deleted
```

Hopefully, you now have a better idea of how to work with a Kubernetes service account using what you learned in this section. Now, let's take a look at how to organize the cluster access using `kubeconfig`.

Organizing the cluster access using kubeconfig

As a Kubernetes user, when you deploy the Kubernetes cluster with `kubeadm`, you will find a file called `config` in the `$HOME/.kube` directory:

```
cloudmelon@cloudmelonplayground:~$ cd $HOME/.kube
cloudmelon@cloudmelonplayground:~/.kube$ ls
cache/  config
```

In other cases, this `kubeconfig` file can be set up as a KUBECONFIG environment variable or a `--kubeconfig` flag. You can find detailed instructions in the official documentation: `https://kubernetes.io/docs/tasks/access-application-cluster/configure-access-multiple-clusters/`.

The `kubeconfig` files help organize information clusters, users, and namespaces. From the `kubectl` utility point of view, it reads `kubeconfig` files to locate the information of the cluster and communicate with the API server of that Kubernetes cluster.

The following is an example of a `kubeconfig` file:

```
apiVersion: v1
clusters:
- cluster:
    certificate-authority: /home/cloudmelon/.minikube/ca.crt
    extensions:
    - extension:
        last-update: Wed, 11 May 2022 23:47:43 UTC
        provider: minikube.sigs.k8s.io
        version: v1.25.2
      name: cluster_info
    server: https://192.168.49.2:8443
  name: minikube
contexts:
- context:
    cluster: minikube
    extensions:
    - extension:
        last-update: Wed, 11 May 2022 23:47:43 UTC
        provider: minikube.sigs.k8s.io
        version: v1.25.2
      name: context_info
    namespace: default
    user: minikube
  name: minikube
current-context: minikube
kind: Config
preferences: {}
users:
```

```
- name: minikube
  user:
    client-certificate: /home/cloudmelon/.minikube/profiles/
minikube/client.crt
    client-key: /home/cloudmelon/.minikube/profiles/minikube/
client.key
```

You can see `config` by using the following command:

```
kubectl config view
```

The output should look as follows:

```
[cloudmelon@cloudmelonplayground:~/.kube$ kubectl config view
apiVersion: v1
clusters:
- cluster:
    certificate-authority: /home/cloudmelon/.minikube/ca.crt
    extensions:
    - extension:
        last-update: Wed, 11 May 2022 23:47:43 UTC
        provider: minikube.sigs.k8s.io
        version: v1.25.2
      name: cluster_info
    server: https://192.168.49.2:8443
  name: minikube
contexts:
- context:
    cluster: minikube
    extensions:
    - extension:
        last-update: Wed, 11 May 2022 23:47:43 UTC
        provider: minikube.sigs.k8s.io
        version: v1.25.2
      name: context_info
    namespace: default
    user: minikube
  name: minikube
current-context: minikube
kind: Config
preferences: {}
users:
- name: minikube
  user:
    client-certificate: /home/cloudmelon/.minikube/profiles/minikube/client.crt
    client-key: /home/cloudmelon/.minikube/profiles/minikube/client.key
```

Figure 6.4 – The kubectl config view output

You can use the `kubectl config` command to display `current-context`:

```
kubectl config current-context
```

The returned output will be the current context – in my case, it is `minikube`. You may notice that it is the same as `current-context` shown in the aforementioned `config` file:

```
minikube
```

To know more about how to organize the cluster access using `kubeconfig`, refer to the official article to learn more:

```
https://kubernetes.io/docs/concepts/configuration/organize-cluster-access-kubeconfig/
```

Configuring access to multiple clusters

As a Kubernetes user, when it comes to multiple clusters, we can also use the `kubectl config` command to configure the current context to switch between different Kubernetes clusters. To find all the commands provided by `kubectl config`, use this command:

```
kubectl config --help
```

The following is an example of how `kubeconfig` contains the access information of two different Kubernetes clusters:

```
apiVersion: v1
clusters:
- cluster:
    certificate-authority-data:
 < authority data >
    server: https://xx.xx.xx.xx
  name: gke_cluster
- cluster:
    certificate-authority-data:
 < authority data >
    server: https://xx.xx.xx.xx
  name: arctestaks
contexts:
- context:
    cluster: gke_cluster
```

```
      user: gke_cluster
    name: gke_cluster
  - context:
      cluster: arctestaks
      user: clusterUser_akscluster
    name: akscluster
  current-context: akscluster
  kind: Config
  preferences: {}
  users:
  - name: gke_cluster
    user:
      auth-provider:
        config:
          access-token:
   < token data >
          cmd-args: config config-helper --format=json
          cmd-path: C:\Program Files (x86)\Google\Cloud SDK\
  google-cloud-sdk\bin\gcloud.cmd
          expiry: '2022-05-12T00:28:06Z'
          expiry-key: '{.credential.token_expiry}'
          token-key: '{.credential.access_token}'
        name: gcp
  - name: clusterUser_akscluster
    user:
      client-certificate-data: <data>
      client-key-data: <data>
      token:
   < token >
```

We could use the `kubectl config current-context` command to see the cluster that I am working on and it would be displayed as the following:

```
gke-cluster
```

The preceding output indicates that I am on a Kubernetes cluster called `gke-cluster` and that I am using the following command to switch my default context to another Kubernetes cluster called `akscluster`:

```
kubectl config use-context akscluster
```

We could use the `kubectl config current-context` command to check my current working Kubernetes cluster and it would be displayed as the following:

```
aks-cluster
```

Switching context is an important technique that you can apply during your actual CKA exam and it's important to perform tasks in the targeting Kubernetes cluster so that you'll be scored accurately. It also comes in handy in your real life working as a Kubernetes administrator, as often, you'll be working on multiple Kubernetes clusters.

To know more about how to configure access to multiple clusters, check the official article: `https://kubernetes.io/docs/tasks/access-application-cluster/configure-access-multiple-clusters/`

Kubernetes authorization

In Kubernetes, a request must be authenticated before it can be authorized with permissions granted to access the Kubernetes cluster resources.

There are four authorization modes in Kubernetes:

- **RBAC authorization**: Kubernetes RBAC is more about regulating access to Kubernetes resources according to the roles with specific permissions to perform a specific task, such as reading, creating, or modifying through an API request. We'll focus on Kubernetes RBAC in this section.

- **Node authorization**: As the name suggests, this grants permissions to the API requests made by `kubelets agent`. This is a special - purpose authorization mode not covered in the CKA exam. You can check out the official documentation about node authorization to find out more: `https://kubernetes.io/docs/reference/access-authn-authz/node/`.

- **ABAC authorization**: ABAC is an access control granted to users by policies and attributes such as user attributes, resource attributes, and objects. This topic is not covered in the current CKA exam. If you want to learn more about using the ABAC mode, you can refer to the official article: `https://kubernetes.io/docs/reference/access-authn-authz/abac/`.

- **Webhook authorization**: Webhook authorization through WebHooks is an HTTP POST triggered by an event. An example of this is that the Webhook will react to a URL when triggered by certain actions. This topic is not covered in the current CKA exam. You can explore the following article if you want to know more about it: `https://kubernetes.io/docs/reference/access-authn-authz/webhook/`.

Let's take a look at what the key areas covered in the CKA exam are, starting with Kubernetes RBAC.

Kubernetes RBAC

Kubernetes RBAC aims to regulate access to Kubernetes resources according to the roles with specific permissions to perform a specific task.

Once specified, RBAC checks the `rbac.authorization.k8s.io` API group membership to see whether it is allowed through the Kubernetes API.

Let's take a look at the different Roles and RoleBindings in Kubernetes.

Roles versus ClusterRoles and their RoleBindings

In Kubernetes, we have Roles and ClusterRoles. A Kubernetes RBAC Role or ClusterRole represents a role with a set of permissions. In a nutshell, they differ by the scope of these permissions:

- A **Role** represents permissions within a particular namespace

- A **ClusterRole** represents permissions within the cluster – it could be cluster-wide, across multiple namespaces, or individual namespaces

With Roles and ClusterRoles, we have the concept of **RoleBinding** and **ClusterRoleBinding**. The bindings bind the role to a list of subjects such as users, groups, or service accounts, as can be seen in the following figure:

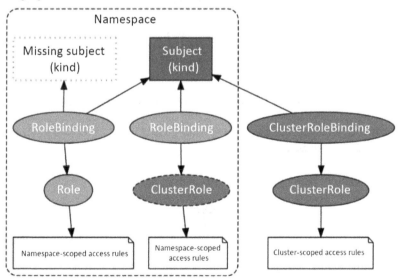

Figure 6.5 – Kubernetes RBAC

Let's define a new role called `dev-user` in a namespace called `dev`. We can use the following command to do this:

```
kubectl create role dev-user --verb=get --verb=list
--resource=pods --namespace=dev
```

The preceding command is the same as the following YAML definition:

```
apiVersion: rbac.authorization.k8s.io/v1
kind: Role
metadata:
  namespace: dev
  name: dev-user
rules:
- apiGroups: [""]
  resources: ["pods"]
  verbs: ["get", "list"]
```

The output of the preceding command is the following:

```
role.rbac.authorization.k8s.io/dev-user created
```

Then, we can use the `kubectl get role` command to check the role that we have just created:

```
cloudmelon@cloudmelonplayground:~$ k get role -n dev
NAME         CREATED AT
dev-user     2022-05-13T04:14:59Z
```

We then need to create the RoleBinding to bind this role to the subjects as follows:

```
kubectl create rolebinding dev-pods-binding --role=dev-user -
-user=melon-dev --namespace=dev
```

Alternatively, we could also use the following YAML file:

```
apiVersion: rbac.authorization.k8s.io/v1
kind: RoleBinding
metadata:
  name: dev-pods-binding
  namespace: dev
subjects:
- kind: User
```

```
   apiGroup: rbac.authorization.k8s.io
   name:melon-dev
 roleRef:
   kind: Role
   name: dev-user
   apiGroup: rbac.authorization.k8s.io
```

Let's define a new ClusterRole called secret-reader – note that the ClusterRole is not namespaced. We could use the following YAML definition:

```
apiVersion: rbac.authorization.k8s.io/v1
kind: ClusterRole
metadata:
  name: secret-reader
rules:
- apiGroups: [""]
  resources: ["secrets"]
  verbs: ["get", "list"]
```

Then, we need to create the RoleBinding to bind this role to the subjects, as shown in the following YAML definition:

```
apiVersion: rbac.authorization.k8s.io/v1
kind: RoleBinding
metadata:
  name: read-secrets
  namespace: development
subjects:
- kind: Group
  name: manager
  apiGroup: rbac.authorization.k8s.io
roleRef:
  kind: ClusterRole
  name: secret-reader
  apiGroup: rbac.authorization.k8s.io
```

We can use the following command to get all the roles across all the namespaces:

```
cloudmelon@cloudmelonplayground:~$ kubectl get roles -A
NAMESPACE            NAME                                      CREATED
AT
dev                  dev-user
kube-public          kubeadm:bootstrap-signer-clusterinfo
kube-public          system:controller:bootstrap-signer
kube-system          extension-apiserver-authentication-reader
kube-system          kube-proxy
kube-system          kubeadm:kubelet-config-1.23
kube-system          kubeadm:nodes-kubeadm-config
kube-system          system::leader-locking-kube-controller-
manager
kube-system          system::leader-locking-kube-scheduler
kube-system          system:controller:bootstrap-signer
kube-system          system:controller:cloud-provider
kube-system          system:controller:token-cleaner
kube-system          system:persistent-volume-provisioner
```

We can use the following command to get all the RoleBindings across all the namespaces:

```
cloudmelon@cloudmelonplayground:~$ kubectl get rolebindings -A
NAMESPACE            NAME
ROLE                                                         AGE
dev                  dev-pods-binding
Role/dev-user                                                15s
kube-public          kubeadm:bootstrap-signer-clusterinfo
Role/kubeadm:bootstrap-signer-clusterinfo         6d
kube-public          system:controller:bootstrap-signer
Role/system:controller:bootstrap-signer           6d
kube-system          kube-proxy
Role/kube-proxy                                   6d
kube-system          kubeadm:kubelet-config-1.23
Role/kubeadm:kubelet-config-1.23                  6d
kube-system          kubeadm:nodes-kubeadm-config
Role/kubeadm:nodes-kubeadm-config                 6d
kube-system          metrics-server-auth-reader
Role/extension-apiserver-authentication-reader    3h
```

```
kube-system          system::extension-apiserver-authentication-
reader    Role/extension-apiserver-authentication-reader
6d
kube-system          system::leader-locking-kube-controller-
manager        Role/system::leader-locking-kube-controller-
manager    6d
kube-system          system::leader-locking-kube-scheduler
Role/system::leader-locking-kube-scheduler              6d
kube-system          system:controller:bootstrap-signer
Role/system:controller:bootstrap-signer                 6d
kube-system          system:controller:cloud-provider
Role/system:controller:cloud-provider                   6d
kube-system          system:controller:token-cleaner
Role/system:controller:token-cleaner                    6d
kube-system          system:persistent-volume-provisioner
Role/system:persistent-volume-provisioner               6d
```

Knowing the ways that Roles and RoleBindings work in Kubernetes, let's now take a look at how to implement your own Kubernetes RBAC Roles and RoleBindings.

Implementing Kubernetes RBAC

To enable RBAC, set `apiserver -authorization-mode` to RBAC, which defaults to `AlwaysAllow`. The other possible values include `node`, `ABAC`, `Always deny`, and webhook. In the following command, we're showing an example of setting it to use Kubernetes RBAC:

```
kube-apiserver -authorization-mode=RBAC
```

To know more about how to set up authorization mode, visit the following link: `https://kubernetes.io/docs/reference/command-line-tools-reference/kube-apiserver/`

Let's start by creating a new deployment using our current context, which is `minikube`:

```
kubectl create deployment mybusybox --image=busybox
```

Then, switch to the context for `dev-user`:

```
kubectl config use-context dev-user
```

As our `dev-user` only has `list` and `get` permissions, let's try to use this profile to delete the deployment:

```
cloudmelon@cloudmelonplayground:~$ kubectl delete deployment
mybusybox
Error from server (Forbidden): deployments.apps is forbidden:
User "dev-user" cannot delete resource "deployments" in API
group "apps" in the namespace "t
```

Now that we have learned how to manage our own Kubernetes RBAC roles, let's take a look at how to manage the security of Kubernetes applications.

Managing the security of Kubernetes applications

A `securityContext` field defines the privilege and access control settings for a Pod in the Pod YAML specification. We need to configure the security context in case a Pod or container needs to interact with the security mechanisms of the underlying operating system unconventionally, and in this section, we'll introduce how to configure a security context for a Pod or container.

As a part of your prep work, you can create a new user and a new group as shown in the following:

```
sudo useradd -u 2000 container-user-0
sudo groupadd -g 3000 container-group-0
```

We will now log in to the worker node and create a new `.txt` file called `message.txt`:

```
sudo mkdir -p /etc/message
echo "hello Packt" | sudo tee -a /etc/message/message.txt
```

From here, you'll see the message that we input from the terminal:

```
hello Packt
```

Now, we need to adjust the permission to limit the permission for testing purposes, which is shown as the following:

```
sudo chown 2000:3000 /etc/message/message.txt
sudo chmod 640 /etc/message/message.txt
```

Finally, we could deploy a new Pod in our current Kubernetes cluster to test it out. The securityContext field is defined as part of a Pod's YAML spec called pod-permission.yaml. With a section called securityContext, we can specify the security permissions information, as in the following YAML file:

```
apiVersion: v1
kind: Pod
metadata:
  name: melon-securitycontext-pod
spec:
  securityContext:
    runAsUser: 2000
    fsGroup: 3000
  containers:
  - name: melonapp-secret-container
    image: busybox
    command: ['sh', '-c','cat /message/message.txt && sleep 3600']
    volumeMounts:
    - name: message-volume
      mountPath: /message
  volumes:
  - name: message-volume
    hostPath:
      path: /etc/message
```

In the preceding YAML definition file, the runAsUser field means that for any container in this Pod, all processes run with a user ID of 2000. The fsGroup field is 2000, which means that all the processes of the container are also part of the supplementary group, ID 2000. The owner for volume/message and any files created in that volume will be the ID 2000 group.

Let's go ahead and deploy this YAML file as follows:

```
kubectl apply -f pod-permission.yaml
```

Then, we'll see the Pod is spinning up but will quickly encounter the following error:

```
NAME                           READY    STATUS             RESTARTS
AGE
melon-securitycontext-pod 0/1         CrashLoopBackOff    1 5m
```

From the preceding example, we can see the Pod is `BackOff` due to the lack of permission. Now, let's pull a similar example to see whether we can fix this. Let's configure a YAML file with a similar configuration to the following:

```
securityContext:
    runAsUser: 1000
    runAsGroup: 3000
    fsGroup: 2000
```

Let's deploy this using the following YAML example:

```
apiVersion: v1
kind: Pod
metadata:
  name: security-context-message
spec:
  securityContext:
    runAsUser: 1000
    runAsGroup: 3000
    fsGroup: 2000
  volumes:
  - name: sec-ctx-msg
    emptyDir: {}
  containers:
  - name: sec-ctx-msg
    image: busybox:1.28
    command: [ "sh", "-c", "sleep 1h" ]
    volumeMounts:
    - name: sec-ctx-msg
      mountPath: /message
    securityContext:
      allowPrivilegeEscalation: false
```

We can see this example is now up and running in my local Kubernetes cluster:

```
cloudmelon@cloudmelonplayground:/$ kubectl get pod security-
context-demo
NAME                          READY    STATUS    RESTARTS    AGE
security-context-message    1/1      Running    0              3m4s
```

Let's get inside this running pod:

```
kubectl exec -it security-context-message -- sh
```

Then, we'll get into the interactive shell, input id, and we'll get the following output:

```
/ $ id
uid=1000 gid=3000 groups=2000
```

From the output, we can see that uid is 1000, the same as the runAsUser field; the gid is 3000, the same as the runAsGroup field; and the fsGroup is 2000.

To learn more about the security context, check out the official documentation here: https://kubernetes.io/docs/tasks/configure-pod-container/security-context/

Summary

This chapter gave an overview of Kubernetes security with a focus on three key topics about container security, RBAC, and the security context. You can use this chapter to assist you with laying the foundations for your CKS exam. With the addition of the next chapter, *Demystifying Kubernetes Networking*, you will get a complete view of working with Kubernetes networking security-related concepts and practice examples to help in your daily work as a Kubernetes administrator, and this will all cover 20% of the CKA exam content. Let's stay tuned!

Mock CKA scenario-based practice test

You have two virtual machines, master-0 and worker-0 – please complete the following mock scenarios.

Scenario 1

Create a new service account named packt-sa in a new namespace called packt-ns.

Scenario 2

Create a Role named `packtrole` and bind it with the RoleBinding `packt-clusterbinding`. Map the `packt-sa` service account with `list` and `get` permissions.

Scenario 3

Create a new pod named `packt-pod` with the `busybox:1.28` image in the `packt-ns` namespace. Expose port `80`. Then, assign the `packt-sa` service account to the Pod.

You can find all the scenario resolutions in *Appendix - Mock CKA scenario-based practice test resolutions* of this book.

FAQs

- *Where can I find the latest updates about Kubernetes security while working with Kubernetes?*

  The Kubernetes Security **Special Interest Group** (**SIG**) has a GitHub repository, which you can find here: `https://github.com/kubernetes/community/tree/master/sig-security`.

- *What is the recommended Kubernetes official article for configuring the ephemeral storage?*

  I recommend bookmarking the official documentation about Kubernetes RBAC, which you can find here: `https://kubernetes.io/docs/reference/access-authn-authz/rbac/`.

7

Demystifying Kubernetes Networking

This chapter will use the Kubernetes networking model to describe some core concepts, as well as how to configure Kubernetes networking on the cluster nodes and network policies. We will also learn about how to configure Ingress controllers and Ingress resources, how to configure and leverage CoreDNS, and how to choose an appropriate container network interface plugin. This content covered in this chapter makes up about 20% of the CKA exam.

In this chapter, we're going to cover the following topics:

- Understanding the Kubernetes networking model
- Configuring Kubernetes networking on the cluster nodes
- Configuring network policies
- Configuring Ingress controllers and Ingress resources
- Configuring and leveraging CoreDNS
- Choosing an appropriate container network interface plugin

Technical requirements

To get started, we need to make sure your local machine meets the following technical requirements:

- A compatible Linux host. We recommend a Debian-based Linux distribution such as Ubuntu 18.04 or later.
- Make sure your host machine has at least 2 GB RAM, 2 CPU cores, and about 20 GB of free disk space.

Understanding the Kubernetes networking model

Kubernetes is designed to facilitate the desired state management to host containerized workloads – these workloads take advantage of sharable compute resources. Kubernetes networking resolves the challenge of how to allow different Kubernetes components to communicate with each other and applications on Kubernetes to communicate with other applications, as well as the services outside of the Kubernetes cluster.

Hence, the official documentation summarizes those networking challenges as container-to-container, pod-to-pod, pod-to-service, external-to-service, and node-to-node communications. Now, we are going to break them down one-by-one in this section.

Container-to-container communication

Container-to-container communication mainly refers to the communication between containers inside a pod – a multi-container pod is a good example of this. A multi-container pod is a pod that contains multiple containers and is seen as a single unit. Within a pod, every container shares the networking, which includes the IP address and network ports so that those containers can communicate with one another through `localhost` or standard **inter-process communications** (**IPC**) such as SystemV semaphores or POSIX shared memory. All listening ports are accessible to other containers in the pod even if they're not exposed outside the pod.

The following figure shows how those containers share a local network with each other inside the same pod:

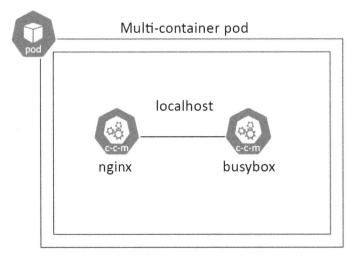

Figure 7.1 – Multiple containers sharing the pod networking

The following is an example called `multi-container-pod.yaml` that shows how to create multi-containers in a pod. In this pod, it contains `nginx` and `busybox` – two containers where `busybox` is a sidecar container that calls `nginx` through port 80 on `localhost`:

```
apiVersion: v1
kind: Pod
metadata:
  name: multi-container-pod
  labels:
      app: multi-container
spec:
  containers:
  - name: nginx
    image: nginx:latest
    ports:
    - containerPort: 80
  - name: busybox-sidecar
    image: busybox:latest
    command: ['sh', '-c', 'while true; do sleep 3600; done;']
```

Let's deploy this `yaml` file by using the `kubectl apply -f multi-container-pod.yaml` command, and the following shows the pod has been created:

```
pod/multi-container-pod created
```

We can use the following command to check whether we could talk to the `nginx` container from the sidecar `busybox` container:

```
kubectl exec multi-container-pod -c busybox-sidecar -- wget
http://localhost:80
```

The following output proves that both containers can talk to each other:

```
cloudmelon@cloudmelonplayground:~$ kubectl exec multi-container-pod -c busybox-sidecar -- wget http://localhost:80
Connecting to localhost:80 (127.0.0.1:80)
saving to 'index.html'
index.html           100% |********************************|   615  0:00:00 ETA
'index.html' saved
```

Figure 7.2 – Connecting to the nginx container from the busybox sidecar

> **Important Note**
>
> A quicker way to create a single container pod by command is by using the following command:
>
> ```
> kubectl run nginx --image=nginx:latest --port=80
> ```
>
> Then, you can use the `kubectl get pods -o yaml` command to export the YAML content, and edit the `yaml` file to add another container.

To double-check that we did indeed get the `nginx` main page from the `busybox` sidecar container, we will use the following command:

```
kubectl exec multi-container-pod -c busybox-sidecar -- cat
index.html
```

The output should look similar to what is shown in *Figure 7.3*:

```
cloudmelon@cloudmelonplayground:~$ kubectl exec multi-container-pod -c busybox-sidecar -- cat index.html
<!DOCTYPE html>
<html>
<head>
<title>Welcome to nginx!</title>
<style>
html { color-scheme: light dark; }
body { width: 35em; margin: 0 auto;
font-family: Tahoma, Verdana, Arial, sans-serif; }
</style>
</head>
<body>
<h1>Welcome to nginx!</h1>
<p>If you see this page, the nginx web server is successfully installed and
working. Further configuration is required.</p>

<p>For online documentation and support please refer to
<a href="http://nginx.org/">nginx.org</a>.<br/>
Commercial support is available at
<a href="http://nginx.com/">nginx.com</a>.</p>

<p><em>Thank you for using nginx.</em></p>
</body>
</html>
```

Figure 7.3 – Checking out the downloaded html page in the busybox container

To learn more about multi-container pods to see how those containers share storage and networking, refer to *Chapter 4, Application Scheduling and Lifecycle Management*.

Pod-to-pod communication

In Kubernetes, each pod has been given a unique IP address based on the `podCIDR` range of that worker node. Although this IP assignment is not permanent, as the pod eventually fails or restarts, the new pod will be assigned a new IP address. By default, pods can communicate with all pods on all nodes through pod networking without setting up **Network Address Translation (NAT)**. This is also where we set up host networking. All pods can communicate with each other without NAT.

Let's deploy a `nginx` pod by using the following command:

```
kubectl run nginx --image=nginx --port=8080
```

The following output shows the pod has been created:

```
pod/nginx created
```

To verify whether the pod has been assigned an IP address, you can use the `kubectl get pod nginx -o wide` command to check the IP address of the `nginx` pod. The output is similar to the following:

```
  NAME      READY    STATUS                  RESTARTS    AGE    IP
  NODE          NOMINATED NODE      READINESS GATES
  nginx     1/1      running    0                 34s    172.17.0.4
  minikube      <none>              <none>
```

You can use the following command to check all pods available in the default namespace and their assigned IP addresses:

```
k get pods -o wide
```

Notice the `IP` column in the following output – it indicates an IP address of `172.17.0.3` for the `multi-container-pod` pod and `172.17.0.4` for the `nginx` pod. These IP addresses assigned to those pods are in the same `podCIDR`:

```
cloudmelon@cloudmelonplayground:~$ k get pods -o wide
NAME                 READY   STATUS    RESTARTS   AGE   IP           NODE       NOMINATED NODE   READINESS GATES
multi-container-pod  2/2     Running   0          38m   172.17.0.3   minikube   <none>           <none>
nginx                1/1     Running   0          67s   172.17.0.4   minikube   <none>           <none>
```

Figure 7.4 – Checking out the IP addresses of the pods

The preceding screenshot also indicates that both pods are on the same node, `minikube`, according to the `NODE` column. We could check the `podCIDR` assigned to the pod by using the following command:

```
kubectl get node minikube -o json | jq .spec.podCIDR
```

The output, which looks as follows, shows the `podCIDR`:

```
10.244.0.0/24
```

From the preceding command output, we can see it does not have the same CIDR as the pods. That's because we tested on a `minikube` cluster. When we start a vanilla `minikube` installation with the `minikube start` command without specifying additional parameters for the CNI network plugin, it sets the default value as `auto`. It chooses a `kindnet` plugin to use, which creates a bridge and then adds the host and the container to it. We'll get to know how to set up a CNI plugin and network policy later in this chapter. To get to know more about `kindnet`, visit the following link: `https://github.com/aojea/kindnet`.

Kubernetes components such as system daemons and `kubelet` can communicate with all pods on the same node. Understanding the connectivity between pods is required for the CKA exam. You can check out the official documentation about cluster networking if you want to learn more here: `https://kubernetes.io/docs/concepts/cluster-administration/networking/#the-kubernetes-network-model`.

Pod-to-service and external-to-service communications

Effective communication between pods and services entails letting the service expose an application running on a set of pods. The service accepts traffic from both inside and outside of the cluster. The set of pods can load - balance across them – each pod is assigned its own IP address and a single DNS.

Similar to pod-to-service, the challenge with external-to-service communication challenge is also resolved by the service. Service types such as a `NodePort` or a `LoadBalancer` can receive traffic from outside the Kubernetes cluster.

Let's now take a look at different *service types* and *endpoints*.

An overview of Kubernetes service types

There are a few types of publishing services in the Kubernetes networking space that are very important. This is different from a headless service. You can visit this link if you want to learn about headless services, which is out of the scope of the CKA exam: `https://kubernetes.io/docs/concepts/services-networking/service/#headless-services`.

The following are the most important types of publishing services that frequently appear in the CKA exam:

Service type	Description	Example
ClusterIP	A default service type for Kubernetes. For internal communications, exposing the service makes it reachable within the cluster.	Checking out the pod address by using the `kubectl get pod mypod -o wide` – the internal IP is `172.17.0.4`
NodePort	For both internal and external communication. `NodePort` exposes the service on a static port on each worker node – meanwhile, a `ClusterIP` is created for it, and it is used for internal communication, requesting the IP address of the node with an open port – for example, `<nodeIP>:<port>` for external communication.	Connecting to a worker node VM with the public IP address `192.0.2.0` from port `80`
LoadBalancer	This works for cloud providers, as it's backed by their respective load balancer offerings. Underneath `LoadBalancer`, `ClusterIP` and `NodePort` are created, which are used for internal and external communication.	Checking out the services for a Kubernetes distribution from a cloud provider such as **Azure Kubernetes Service (AKS)** or **Google Kubernetes Engine (GKE)** by using `kubectl get service mysvc -n mynamespace` – the internal IP is `172.17.0.4`
ExternalName	Maps the service to the contents with a CNAME record with its value. It allows external traffic access through it.	For example, `my.packt.example.com`

To learn more about the differences between publishing services and headless services, check here: `https://kubernetes.io/docs/concepts/services-networking/service/#publishing-services-service-types`. Now, let's take a look at each of those services in this section.

ClusterIP

ClusterIP is the default Kubernetes service type for internal communications. In the case of a pod or ClusterIP, the pod is reachable inside the Kubernetes cluster. However, it is still possible to allow external traffic to access the ClusterIP via kube-proxy, which creates iptables entries. It comes in handy in some use cases, such as displaying Kubernetes dashboards. *Figure 7.5* describes how the network traffic load - balances (round-robin) and routes to the pod. Then, it goes through ClusterIP or other services before hitting the pods:

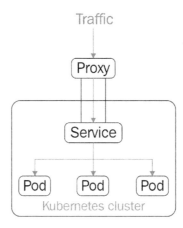

Figure 7.5 – ClusterIP and kube-proxy

Through the preceding diagram, we get a first look at how the service works with the pods. Let's go ahead and deploy an application and do a deeper dive. To create a deployment called nginx and with the replicas number of 2, use the following command:

```
kubectl create deployment nginx --image=nginx --replicas=2
```

We can track down the process of deployment by the following command:

```
kubectl get deploy nginx -o wide
```

Once we do, we should be able to see the following output:

NAME	READY	UP-TO-DATE	AVAILABLE	AGE	CONTAINERS	IMAGES	SELECTOR
nginx	2/2	2	2	3m6s	nginx	nginx	app=nginx

Figure 7.6 – The available nginx replica counts

From the preceding output, we can see that two copies of the nginx pod are up and running, just to get a better understanding of those pods. We can see how those nginx pods are presented in the default namespace.

Note that we're doing the test in the default namespace for simplicity. You can add the -n flag to work with deployment and pods in a different namespace. Refer to *Chapter 4, Application Scheduling and Lifecycle Management,* to see how the application deployment in Kubernetes works. Go and try the following command:

```
kubectl get pods
```

The output will return all the available pods in the default namespace:

```
NAME                     READY   STATUS    RESTARTS   AGE
nginx                    1/1     Running   0          18h
nginx-8f458dc5b-p74rr    1/1     Running   0          4m37s
nginx-8f458dc5b-v8j74    1/1     Running   0          4m37s
```

Figure 7.7 – The available nginx pods in the default namespace

Now, we're exposing these pods to the Kubernetes cluster. We're using the following command to create a service called melon-service:

```
kubectl expose deployment nginx --type=ClusterIP --port 8080
--name=melon-service --target-port 80
```

From the preceding command, we can see that we have created a ClusterIP type of service. We can specify the following flags:

- type is the type of service – in our case, it is ClusterIP. We'll take a look at NodePort and LoadBalancer in the next sections of this chapter.

- port is the port that the service serves on.

- target-port is the port on the container to which the service redirects the traffic.

> **Important Note**
>
> Understanding those command flags will help you use them smoothly; I recommend remembering this command so that you can quickly recall it during the actual CKA exam. You can also refer to the following link (https://kubernetes.io/docs/reference/generated/kubectl/kubectl-commands#expose) to understand whether other flags will help you along the way.

The output of the previous command should look similar to the following:

```
service/melon-service exposed
```

The preceding command is executed successfully based on this output. Now, let's go to the default namespace and check out all the available services using the `kubectl get svc` command – this will give you the following output:

```
NAME            TYPE        CLUSTER-IP      EXTERNAL-IP   PORT(S)    AGE
kubernetes      ClusterIP   10.96.0.1       <none>        443/TCP    28h
melon-service   ClusterIP   10.102.194.57   <none>        8080/TCP   7s
```

Figure 7.8 – The available nginx pods in the default namespace

The preceding output shows the `ClusterIP` type has been created with an IP address of `10.102.194.57` and this service serves on a port of `8080`.

What we did in this section to create a new `ClusterIP` service by using the `kubectl expose` command can also be done using the following YAML manifest file:

```
apiVersion: v1
kind: Service
metadata:
  name: melon-service
spec:
  type: ClusterIP
  selector:
    app: nginx
  ports:
  - protocol: TCP
    port: 8080
    targetPort: 80
```

From the preceding YAML definition, we can see there's a section called `selector`. This section has a key-value pair, `app:nginx`, that has a label sector. Usually, we use a selector to map the service with the pods. Here's the YAML definition of the `nginx` deployment if we didn't go for the `kubectl` command:

```
apiVersion: apps/v1
kind: Deployment
metadata:
  name: nginx
spec:
  selector:
    matchLabels:
```

```
      app: nginx
  replicas: 2
  template:
    metadata:
      labels:
        app: nginx
    spec:
      containers:
      - name: nginx
        image: nginx
        ports:
        - containerPort: 80
```

From the preceding YAML definition, we can see that there is a section to specify the selector and we used the same key-value pair, `app: nginx`, to map the `ClusterIP` specification so that it worked as expected. Refer to *Chapter 4, Application Scheduling and Lifecycle Management,* to learn more about label sectors.

> **Important Note**
>
> As we mentioned before, the CKA exam is about time management, so it will be much more efficient to use commands to achieve the goal.

A corresponding endpoints object can achieve what we have discussed without using a selector. You can use the following commands to get the endpoints of `melon-service`:

```
k get ep melon-service
```

The following is the output of the preceding command:

```
NAME            ENDPOINTS                       AGE
melon-service   10.1.0.32:80,10.1.0.33:80       6m9s
```

Figure 7.9 – Display the endpoints of the nginx pods in the default namespace

As you can see, there's nothing specific in the YAML definition file that we defined here. We can compare the service definition by exporting its YAML definition using the following command:

```
kubectl get svc  melon-service -o yaml
```

We will be able to see the exported output as follows:

```yaml
apiVersion: v1
kind: Service
metadata:
  creationTimestamp: "2022-06-12T22:06:18Z"
  labels:
    app: nginx
  name: melon-service
  namespace: default
  resourceVersion: "128419"
  uid: 76d0969d-d211-467b-8952-a4699c7599de
spec:
  clusterIP: 10.102.194.57
  clusterIPs:
  - 10.102.194.57
  internalTrafficPolicy: Cluster
  ipFamilies:
  - IPv4
  ipFamilyPolicy: SingleStack
  ports:
  - port: 8080
    protocol: TCP
    targetPort: 80
  selector:
    app: nginx
  sessionAffinity: None
  type: ClusterIP
status:
  loadBalancer: {}
```

Figure 7.10 – The definition of the nginx service in the default namespace

Comparing this exported definition with what we have walked through in this section using `kubectl` and a YAML definition will help you understand the services in Kubernetes better. Now, let's take a look at another important service in Kubernetes, called `NodePort`.

NodePort

`NodePort` opens ports on the Kubernetes nodes, which usually are de facto virtual machines. `NodePort` exposes access through the IP of the nodes and, with the opened port, makes the application accessible from outside of the Kubernetes cluster. The network traffic is forwarded from the ports to the service. `kube-proxy` allocates a port in the range `30000` to `32767` on every node – it works as shown in the following figure:

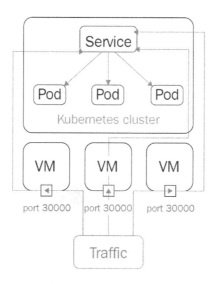

Figure 7.11 – A NodePort in Kubernetes

With the preceding diagram, we get a closer look at how `NodePort` works with the pods. Let's go ahead and create a deployment called `webfront-app` with a `replicas` number of 2 using the following command:

```
kubectl create deployment webfront-app --image=nginx
--replicas=2
```

If it's created successfully, you will see the following output:

```
deployment.apps/webfront-app created
```

Then, we can go ahead and use the following command to expose a web frontend using `NodePort`:

```
kubectl expose deployment webfront-app --port=8080 --target-
port=80 --type=NodePort
```

The following output shows that we have exposed `webfront-app` successfully:

```
service/webfront-app exposed
```

Note that if you don't provide a target port, it is assumed to be the same as the container port. Also note that if you don't provide a node port, a free port in the range between 30000 and 32767 is automatically allocated.

Now, let's check all the services that we have just created. As we didn't specify the name in the previous command, the service name is presumed to be the same as the application name:

```
kubectl get svc webfront-app -o wide
```

The output should look as follows:

```
NAME           TYPE       CLUSTER-IP     EXTERNAL-IP   PORT(S)        AGE   SELECTOR
webfront-app   NodePort   10.97.148.160  <none>        80:31400/TCP   13m   app=webfront-app
```

Figure 7.12 – The webfront-app NodePort in the default namespace

From the preceding output, we can see the port is exposed at 31400, which is in the range of 30000 to 32767 on the node, and the target port is 80, which is opened at the container level. So, let's get the node IP by using the following command:

```
kubectl get node -o wide
```

The key part of your output is as follows:

```
NAME            STATUS   ROLES          AGE   VERSION   INTERNAL-IP
docker-desktop  Ready    control-plane  30h   v1.24.0   192.168.65.4
```

Figure 7.13 – The internal IP of the webfront-app NodePort

From the preceding output, we are getting the internal IP of the node, as we're testing locally, so we can use the internal IP and port in conjunction to connect to webfront-app:

```
192.168.65.4:31400
```

Let's deploy a new nginx pod called sandbox-nginx to test out the connectivity by using the following command:

```
kubectl run -it sandbox --image=nginx --rm --restart=Never --
curl -Is http://192.168.65.4:31400
```

The output is similar to the following:

```
HTTP/1.1 200 OK
Server: nginx/1.21.6
Date: Sun, 12 Jun 2022 23:43:43 GMT
Content-Type: text/html
Content-Length: 615
Last-Modified: Tue, 25 Jan 2022 15:03:52 GMT
Connection: keep-alive
ETag: "61f01158-267"
Accept-Ranges: bytes

pod "sandbox-nginx" deleted
```

Figure 7.14 – The internal IP of the webfront-app NodePort

In the actual CKA exam, you'll be working on a few different VMs. In case you need to connect to the application deployed on that node, you can use the following command to get the external IPs of all nodes:

```
kubectl get nodes -o jsonpath='{.items[*].status.addresses[?(
@.type=="ExternalIP")].address}'
```

Similarly, if you want to get the internal IPs of all nodes, you can use the following command:

```
kubectl get nodes -o jsonpath='{.items[*].status.addresses[?(
@.type==" InternalIP ")].address}'
```

In the actual exam, you can also connect to that node using the internal IP, and then use the following command, which will give you the same result:

```
curl -Is http://192.168.65.4:31400
```

In the case that you have a public IP address of the node VM that you can ping from your local environment, you can use the following command:

```
curl -Is http://<node external IP>:<node port>
```

> **Tips and Tricks**
>
> Some important JSONPath commands can be found on the Kubernetes cheat sheets here if you need some help: https://kubernetes.io/docs/reference/kubectl/cheatsheet/#viewing-finding-resources.

What we did in this section to create a new `NodePort` service by using the `kubectl expose` command can also be done using the following YAML manifest file:

```
apiVersion: v1
kind: Service
metadata:
  name: webfront-app
  labels:
    app: webfront-app
spec:
  ports:
  - port: 8080
    targetPort: 80
  selector:
    app: webfrontapp
  type: NodePort
```

Public cloud providers often support an external load balancer, which we can define as `LoadBalancer` when working with Kubernetes. Now, let's take a look at it in the following section.

LoadBalancer

`LoadBalancer` is a standard way to connect a service from outside of the cluster. In this case, a network load balancer redirects all external traffic to a service, as shown in the following figure, and each service gets its own IP address. It allows the service to load - balance the network traffic across applications:

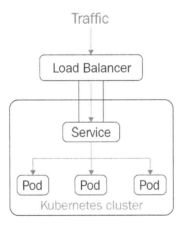

Figure 7.15 – LoadBalancer in Kubernetes

`LoadBalancer` is not a popular topic in the CKA exam, as it only works in a cloud environment or another environment that supports external load balancers. Deploying the `LoadBalancer` service to get a public IP is commonly used in managed Kubernetes distributions such as **Azure Kubernetes Service (AKS)**, **Elastic Kubernetes Service (EKS)**, and **Google Kubernetes Engine (GKE)**. `LoadBalancer` is the default outbound type for AKS – the following is a sample YAML definition in that regard:

```
apiVersion: v1
kind: Service
metadata:
  name: packt-svc
spec:
  type: LoadBalancer
  ports:
  - port: 80
    targetPort: 8080
  selector:
    app: my-packt-app
```

We could also use the `kubectl expose` command to do so:

```
kubectl expose deployment nginx --port=80 --target-port=8080 \
        --name=packt-svc --type=LoadBalancer
```

The output of the preceding command is as follows:

```
NAME            TYPE           CLUSTER-IP       EXTERNAL-IP   PORT(S)        AGE
kubernetes      ClusterIP      10.96.0.1        <none>        443/TCP        33h
melon-service   ClusterIP      10.102.194.57    <none>        8080/TCP       4h57m
nginx-svc       ClusterIP      10.107.75.83     <none>        80/TCP         23h
packt-svc       LoadBalancer   10.96.153.242    localhost     80:31055/TCP   42s
webfront-app    NodePort       10.97.148.160    <none>        80:31400/TCP   3h42m
```

Figure 7.16 – LoadBalancer output in Kubernetes

Since I was testing LoadBalancer in Docker Desktop with WSL2, it was not supported – the preceding output shows that `EXTERNAL-IP` is `localhost`. Although, when I was working on AKS, it showed the real public IP address. Refer to this link to see what worked out for me: `https://docs.microsoft.com/en-us/azure/aks/load-balancer-standard`.

ExternalName

`ExternalName` maps the service to the contents with a CNAME record with its value. It allows external traffic to access it. The following is the sample YAML definition for `ExternalName`:

```
apiVersion: v1
kind: Service
metadata:
  name: my-packt-svc
  namespace: prod
spec:
  type: ExternalName
  externalName: my.melonapp.packt.com
```

Note that the preceding `ExternalName` type is defined as `my.melonapp.packt.com` – we could use the `nslookup` command to check `my-packt-svc.default`, `svc.cluster.local`. This returns the CNAME record for `my.melonapp.packt.com`. We'll dive deeper into how the DNS in Kubernetes works later in this chapter.

Check services and endpoints

In this section, we have worked on all four of the common service types in Kubernetes. In case we need to quickly check all the services across all namespaces, we can use the following command:

```
kubectl get services --all-namespaces
```

Alternatively, we can use the following command:

```
kubectl get svcs -A
```

The following shows the output for the preceding command:

NAMESPACE	NAME	TYPE	CLUSTER-IP	EXTERNAL-IP	PORT(S)	AGE
default	kubernetes	ClusterIP	10.96.0.1	\<none\>	443/TCP	33h
default	melon-service	ClusterIP	10.102.194.57	\<none\>	8080/TCP	5h1m
default	nginx-svc	ClusterIP	10.107.75.83	\<none\>	80/TCP	23h
default	packt-svc	LoadBalancer	10.96.153.242	localhost	80:31055/TCP	5m8s
default	webfront-app	NodePort	10.97.148.160	\<none\>	80:31400/TCP	3h46m
kube-system	kube-dns	ClusterIP	10.96.0.10	\<none\>	53/UDP,53/TCP,9153/TCP	33h

Figure 7.17 – Getting all the services across different namespaces

The preceding screenshot lists services across namespaces, as well as their `ClusterIP` and port information. If you want to check out a specific service, you can use the following:

```
kubectl get svc <service-name> -n <namespace>
```

The example of the preceding command is `kubectl get svc kube-dns -n kube-system`, which will give you the service information. You can also go one step further to check the details by using the `kubectl describe svc` command:

```
kubectl describe svc kube-dns -n kube-system
```

The output of the preceding command is as follows:

```
Name:              kube-dns
Namespace:         kube-system
Labels:            k8s-app=kube-dns
                   kubernetes.io/cluster-service=true
                   kubernetes.io/name=CoreDNS
Annotations:       prometheus.io/port: 9153
                   prometheus.io/scrape: true
Selector:          k8s-app=kube-dns
Type:              ClusterIP
IP Family Policy:  SingleStack
IP Families:       IPv4
IP:                10.96.0.10
IPs:               10.96.0.10
Port:              dns   53/UDP
TargetPort:        53/UDP
Endpoints:         10.1.0.2:53,10.1.0.27:53
Port:              dns-tcp   53/TCP
TargetPort:        53/TCP
Endpoints:         10.1.0.2:53,10.1.0.27:53
Port:              metrics   9153/TCP
TargetPort:        9153/TCP
Endpoints:         10.1.0.2:9153,10.1.0.27:9153
Session Affinity:  None
Events:            <none>
```

Figure 7.18 – Checking the service details

For the endpoints, we can use the following command to check the endpoint of the service:

```
kubectl get endpoints melon-service
```

It can also be as follows:

```
NAME             ENDPOINTS                      AGE
melon-service    10.1.0.32:80,10.1.0.33:80      5h7m
```

In case we'd like to check out all the endpoints across the different namespaces, we have the following:

```
kubectl get ep --all-namespaces
```

The output of the preceding command will list all the endpoints across different namespaces:

```
NAMESPACE     NAME                ENDPOINTS                                        AGE
default       kubernetes          192.168.65.4:6443                                33h
default       melon-service       10.1.0.32:80,10.1.0.33:80                        5h8m
default       nginx-svc           10.1.0.9:80                                      24h
default       packt-svc           10.1.0.32:8080,10.1.0.33:8080                    11m
default       webfront-app        10.1.0.34:80,10.1.0.35:80                        3h53m
kube-system   docker.io-hostpath  <none>                                           33h
kube-system   kube-dns            10.1.0.2:53,10.1.0.27:53,10.1.0.2:53 + 3 more... 33h
```

Figure 7.19 – Getting all the endpoints across different namespaces

The same principle also applies to listing all the endpoints by namespace. When you want to check out a specific service, you can use the following:

```
kubectl get ep <service-name> -n <namespace>
```

We have talked about how to work with services and endpoints in Kubernetes, which covers pod-to-service communication. Now, let's get into node-to-node communication in the next section.

Node-to-node communication

Within a cluster, each node is registered by the `kubelet` agent to the master node, and each node is assigned a node IP address so they can communicate with each other.

To verify this, you can use the `kubectl get node -o wide` command to check the internal IP of each node. The output is similar to the following, in which you'll notice an `internal-IP` for the worker node:

```
cloudmelon@cloudmelonplayground:~$ k get node -o wide
NAME      STATUS  ROLES                AGE  VERSION  INTERNAL-IP   EXTERNAL-IP  OS-IMAGE         KERNEL-VERSION    CONTAINER-RUNTIME
minikube  Ready   control-plane,master 77d  v1.23.3  192.168.49.2  <none>       Ubuntu 20.04.2 LTS  5.4.0-100-generic  docker://20.10.12
```

Figure 7.20 – Checking out the node IP and further information

From the preceding screenshot, we can see the internal IP of the current node is 192.168.49.2. In the case that we have multiple nodes, we can ping each node from the node within the same network. We need to ensure the connectivity between master nodes and worker nodes, so the workloads get to be scheduled to the worker node. In this regard, a good understanding of how to configure the hosting network for Kubernetes nodes is very important. So, let's have a look at the container network interface plugin next.

Choosing an appropriate Container Network Interface plugin

In *Chapter 2, Installing and Configuring Kubernetes Clusters*, we talked about how to use the Calico plugin as the overlay network for our Kubernetes cluster. We can enable the **Container Network Interface** (**CNI**) for pod-to-pod communication. The CNI plugins conform to the CNI specification. Once the CNI is set up on the Kubernetes cluster, it will allocate the IP address per pod.

CNI networking in Kubernetes

There's a wide range of networking plugins working with Kubernetes on today's market, including popular open source frameworks such as Calico, Flannel, Weave Net, and more. For more options, check out the official documentation here: `https://kubernetes.io/docs/concepts/cluster-administration/addons/`.

Taking Flannel as an example, Flannel is focused on configuring a Layer 3 network fabric designed for Kubernetes, mainly for routing packets among different containers. Flannel runs a single binary agent called `flanneld` on each host, which is responsible for allocating a subnet preconfigured address space to each host, as in the following:

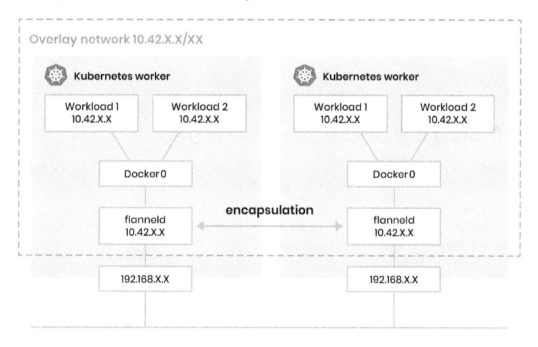

Figure 7.21 – CNI networking in Kubernetes

The preceding figure demonstrates how Flannel CNI networking works. There are many options in the community – let's take a look at the decision metrics about choosing the CNI plugin.

Decision metrics

To make a good choice of an appropriate CNI plugin that fits your requirements, you can refer to the following table of different features from each of the CNI providers mentioned:

	Provider networking	Encapsulation and routing	Support for network policies	Datastore	Encryption	Ingress / Egress
Flannel	Layer 3	VxLAN	No	ETCD	Yes	No
Calico	Layer 3	BGP, eBPF	Yes	ETCD	Yes	Yes
Weavenet	Layer 2	VxLAN	Yes	NO	Yes	Yes
Canal	Layer 2	VxLAN	Yes	ETCD	No	Yes

For quick testing, Flannel is simple to set up. Calico and Weave Net are better options for enterprise-grade customers, as they have a wide range of capabilities. In real life, it is possible to use multiple CNI solutions in a single environment to fulfill some complex networking requirements. However, that's out of reach of the CKA certification exam.

Now let's take a look at the Ingress controller in the next section.

Configuring Ingress controllers and Ingress resources

One of the challenges of Kubernetes networking is about managing internal traffic, which is also known as east-west traffic, and external traffic, which is known as north-south traffic.

There are a few different ways of getting external traffic into a Kubernetes cluster. When it comes to Layer 7 networking, Ingress exposes HTTP and HTTPS at Layer 7 routes from outside the cluster to the services within the cluster.

How Ingress and an Ingress controller works

Ingress acts as a router to route traffic to services via an Ingress-managed load balancer – then, the service distributes the traffic to different pods. From that point of view, the same IP address can be used to expose multiple services. However, our application can become more complex, especially when we need to redirect the traffic to its subdomain or even a wild domain. Ingress is here to address these challenges.

Ingress works with an Ingress controller to evaluate the defined traffic rules and then determine how the traffic is being routed. The process works as shown in *Figure 7.22*:

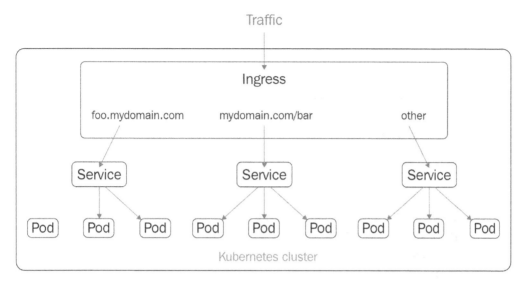

Figure 7.22 – Ingress resources in Kubernetes

In addition to what we see here in *Figure 7.22*, Ingress also provides some key capabilities such as load balancing, SSL termination, and name-based virtual hosting.

We need to deploy an Ingress controller in the Kubernetes cluster and then create Ingress resources. We are using `ingress-nginx` as an example in this section. We have a wide range of options for Ingress controllers on the market nowadays. Check out the official documentation here to get more details: `https://kubernetes.io/docs/concepts/services-networking/ingress-controllers/`.

Using multiple Ingress controllers

Note that it is also possible to deploy multiple Ingress controllers by using the Ingress class within a Kubernetes cluster. Refer to this article to get more details: `https://kubernetes.io/docs/concepts/services-networking/ingress-controllers/#using-multiple-ingress-controllers`.

Work with Ingress resources

As mentioned, the `nginx` Ingress controller is one of the most popular in today's market, so we are using it as the main example in this section. We need to deploy an Ingress controller in the Kubernetes cluster and create Ingress resources.

Here, we are defining a minimal `nginx` resource with the following YAML definition:

```yaml
apiVersion: networking.k8s.io/v1
kind: Ingress
metadata:
  name: minimal-ingress
  annotations:
    nginx.ingress.kubernetes.io/rewrite-target: /
spec:
  ingressClassName: packt-nginx
  rules:
  - http:
      paths:
      - path: /packs
        pathType: Prefix
        backend:
          service:
            name: test
            port:
              number: 80
```

From the preceding YAML definition, we know that the `apiVersion`, `kind`, `metadata`, and `spec` fields are mandatory. Then, we also need an Ingress object, which contains a valid DNS subdomain name.

A default `IngressClass` would look as follows:

```yaml
apiVersion: networking.k8s.io/v1
kind: IngressClass
metadata:
  labels:
    app.kubernetes.io/component: controller
  name: nginx-example
  annotations:
    ingressclass.kubernetes.io/is-default-class: "true"
spec:
  controller: k8s.io/ingress-nginx
```

To learn more about how to work with Ingress, check out the official documentation: `https://kubernetes.io/docs/concepts/services-networking/ingress/`.

Ingress annotations and rewrite-target

You can add Kubernetes annotations to specific Ingress objects so that you can customize their behaviors. These annotation keys and values can only be strings. The following is an example of how to add annotations to Ingress resources using nginx as an example:

```
apiVersion: networking.k8s.io/v1
kind: Ingress
metadata:
  annotations:
    nginx.ingress.kubernetes.io/rewrite-target: /
  name: packt-ingress
spec:
  ingressClassName: nginx
  rules:
  - host: book.packt.com
    http:
      paths:
      - path: /packt-book
        pathType: Prefix
        backend:
          service:
            name: packt-svc
            port:
              number: 80
```

There are many annotations available for nginx – you can check them out by visiting the following page: https://kubernetes.github.io/ingress-nginx/user-guide/nginx-configuration/annotations/.

Different Ingress controllers provide different capabilities, often using annotations and rewrite-target to rewrite the default behavior. You can check out here to learn how to rewrite behaviors for nginx Ingress controllers: https://kubernetes.github.io/ingress-nginx/examples/rewrite/#rewrite-target.

We touched on the domain name and subdomain name in this section. Now, it's a good time to talk about how the DNS domain hostname works in Kubernetes. Let's get right into it in the next section:

Configuring and leveraging CoreDNS

As mentioned earlier in this chapter, nodes, pods, and services are assigned their own IP addresses in the Kubernetes cluster. Kubernetes runs a **Domain Name System** (**DNS**) server implementation that maps the name of the service to its IP address via DNS records. So, you can reach out to the services with a consistent DNS name instead of using its IP address. This comes in very handy in the context of microservices. All microservices running in the current Kubernetes cluster can reference the service name to communicate with each other.

The DNS server mainly supports the following three types of DNS records, which are also the most common ones:

- **A** or **AAAA records** for forward lookups that map a DNS name to an IP address. A record maps a DNS name to an IPv4 address, whereas an AAAA record allows mapping a DNS name to an IPv6 address.

- **SRV records** for port lookups so that connections are established between a service and a hostname.

- **PTR records** for reversing IP address lookups, which is the opposite function of A and AAAA records. It matches IP addresses to a DNS name. For example, a PTR record for an IP address of `172.0. 0.10` would be stored under the `10.0. 0.172.in-addr.arpa` DNS zone.

Knowing these basic DNS concepts will help us get a better understanding of DNS in Kubernetes.

In Kubernetes 1.21, `kubeadm` removed support for `kube-dns` for DNS replication. CoreDNS is now becoming the default DNS service. CoreDNS is an extensible DNS server that can serve as a Kubernetes cluster DNS. It is a **Cloud-Native Computing Foundation** (**CNCF**) graduated project, as it's stable and already has use cases running in a production environment successfully. You can check out the version of CoreDNS installed by `kubeadm` for Kubernetes in the past from here: `https://github.com/coredns/deployment/blob/master/kubernetes/CoreDNS-k8s_version.md`.

If your Kubernetes cluster is not on CoreDNS yet, here is an official end-to-end guide to help you migrate to CoreDNS smoothly and avoid backward - incompatible configuration issues: `https://github.com/coredns/deployment/blob/master/kubernetes/Upgrading_CoreDNS.md`.

Check whether the CoreDNS server is up and running

The Kubernetes DNS server schedules a DNS pod and service on the Kubernetes cluster to check whether the DNS server is up and running on your cluster. To do this, you can simply use the following command:

```
kubectl get pods -n kube-system
```

Normally, you should be able to see an output similar to the following:

```
NAME                                      READY    STATUS     RESTARTS        AGE
coredns-6d4b75cb6d-4xcmf                  1/1      Running    0               82m
coredns-6d4b75cb6d-kj6cq                  1/1      Running    0               82m
etcd-docker-desktop                       1/1      Running    0               82m
kube-apiserver-docker-desktop             1/1      Running    0               82m
kube-controller-manager-docker-desktop    1/1      Running    0               82m
kube-proxy-9rfxs                          1/1      Running    0               82m
kube-scheduler-docker-desktop             1/1      Running    0               82m
storage-provisioner                       1/1      Running    0               82m
vpnkit-controller                         1/1      Running    6 (4m21s ago)   82m
```

Figure 7.23 – When multi-container pods share a network

When you're certain that you're on CoreDNS, you can also use the following command:

```
kubectl get pods -n kube-system | grep coredns
```

The output is similar to the following:

```
  coredns-6d4b75cb6d-4xcmf              1/1       Running    0
82m
  coredns-6d4b75cb6d-kj6cq              1/1       Running    0
82m
```

From the previous output, you may have noticed that we have two replicas of the CoreDNS pod. The intention was to set the default value to two copies for high availability when installing CoreDNS. To prove this, you can check out the CoreDNS deployment settings by using the `kubectl describe` command as follows:

```
kubectl describe deploy coredns -n kube-system
```

The output should look similar to the following:

```
Name:                   coredns
Namespace:              kube-system
CreationTimestamp:      Sat, 11 Jun 2022 10:18:48 -0700
Labels:                 k8s-app=kube-dns
Annotations:            deployment.kubernetes.io/revision: 1
Selector:               k8s-app=kube-dns
Replicas:               2 desired | 2 updated | 2 total | 2 available | 0 unavailable
StrategyType:           RollingUpdate
MinReadySeconds:        0
RollingUpdateStrategy:  1 max unavailable, 25% max surge
Pod Template:
  Labels:             k8s-app=kube-dns
  Service Account:  coredns
  Containers:
   coredns:
    Image:         k8s.gcr.io/coredns/coredns:v1.8.6
    Ports:         53/UDP, 53/TCP, 9153/TCP
    Host Ports:    0/UDP, 0/TCP, 0/TCP
    Args:
      -conf
      /etc/coredns/Corefile
    Limits:
      memory:  170Mi
    Requests:
      cpu:        100m
      memory:     70Mi
    Liveness:     http-get http://:8080/health delay=60s timeout=5s period=10s #success=1 #failure=5
    Readiness:    http-get http://:8181/ready delay=0s timeout=1s period=10s #success=1 #failure=3
    Environment:  <none>
    Mounts:
      /etc/coredns from config-volume (ro)
  Volumes:
   config-volume:
    Type:      ConfigMap (a volume populated by a ConfigMap)
    Name:      coredns
    Optional:  false
  Priority Class Name:  system-cluster-critical
Conditions:
  Type           Status  Reason
  ----           ------  ------
  Available      True    MinimumReplicasAvailable
  Progressing    True    NewReplicaSetAvailable
OldReplicaSets:  <none>
NewReplicaSet:   coredns-6d4b75cb6d (2/2 replicas created)
Events:          <none>
```

Figure 7.24 – When multi-container pods share a network

As it's a deployment, we could use a typical kubectl scale command to scale the CoreDNS deployment out and in. This comes in handy when you want to economize some cluster resources. You can scale it down to one replica using the following command:

```
kubectl scale deploy coredns -n kube-system --replicas=1
```

The output should look as follows:

```
deployment.apps/coredns scaled
```

You can then use the `kubectl get deploy` command to check out the number of replicas currently available in the cluster:

```
NAME       READY   UP-TO-DATE   AVAILABLE   AGE
coredns    1/1 1 1 11h
```

Similarly, when you want it to be more resilient by scheduling more replicas, you can use the following command to get more replicas:

```
kubectl scale deploy coredns -n kube-system --replicas=4
```

Alternatively, we can go back to check the number of the replicas by using the following command:

```
kubectl get pods -n kube-system
```

As the following screenshot shows, we managed to increase the number of replicas of `coredns` from one to four:

```
NAME                                            READY   STATUS    RESTARTS      AGE
coredns-6d4b75cb6d-4h89j                        1/1     Running   0             20s
coredns-6d4b75cb6d-kj6cq                        1/1     Running   0             11h
coredns-6d4b75cb6d-pzptr                        1/1     Running   0             20s
coredns-6d4b75cb6d-v74pr                        1/1     Running   0             20s
etcd-docker-desktop                             1/1     Running   0             11h
kube-apiserver-docker-desktop                   1/1     Running   0             11h
kube-controller-manager-docker-desktop          1/1     Running   0             11h
kube-proxy-9rfxs                                1/1     Running   0             11h
kube-scheduler-docker-desktop                   1/1     Running   0             11h
storage-provisioner                             1/1     Running   0             11h
vpnkit-controller                               1/1     Running   57 (12m ago)  11h
```

Figure 7.25 – When multi-container pods share a network

The previous examples also demonstrate that those four replicas of CoreDNS are identical. We can use the `kubectl describe` command to take a closer look at either of those four `coredns` pods. The following command is an example:

```
k describe pod coredns-6d4b75cb6d-4h89j -n kube-system
```

The output should look as follows:

```
Name:                   coredns-6d4b75cb6d-4h89j
Namespace:              kube-system
Priority:               2000000000
Priority Class Name:    system-cluster-critical
Node:                   docker-desktop/192.168.65.4
Start Time:             Sat, 11 Jun 2022 21:52:57 -0700
Labels:                 k8s-app=kube-dns
                        pod-template-hash=6d4b75cb6d
Annotations:            <none>
Status:                 Running
IP:                     10.1.0.27
IPs:
  IP:                   10.1.0.27
Controlled By:          ReplicaSet/coredns-6d4b75cb6d
Containers:
  coredns:
    Container ID:       docker://f56e1739e30da3345fa1bb59d989d687df6a34edc70cfafe4fb38fbae21ea9a8
    Image:              k8s.gcr.io/coredns/coredns:v1.8.6
    Image ID:           docker://sha256:a4ca41631cc7ac19ce1be3ebf0314ac5f47af7c711f17066006db82ee3b75b03
    Ports:              53/UDP, 53/TCP, 9153/TCP
    Host Ports:         0/UDP, 0/TCP, 0/TCP
    Args:
      -conf
      /etc/coredns/Corefile
    State:              Running
      Started:          Sat, 11 Jun 2022 21:52:58 -0700
    Ready:              True
    Restart Count:      0
    Limits:
      memory:   170Mi
    Requests:
      cpu:      100m
      memory:   70Mi
    Liveness:           http-get http://:8080/health delay=60s timeout=5s period=10s #success=1 #failure=5
    Readiness:          http-get http://:8181/ready delay=0s timeout=1s period=10s #success=1 #failure=3
    Environment:    <none>
    Mounts:
      /etc/coredns from config-volume (ro)
      /var/run/secrets/kubernetes.io/serviceaccount from kube-api-access-6wh5k (ro)
Conditions:
  Type              Status
  Initialized       True
  Ready             True
  ContainersReady   True
  PodScheduled      True
Volumes:
  config-volume:
    Type:       ConfigMap (a volume populated by a ConfigMap)
    Name:       coredns
    Optional:   false
  kube-api-access-6wh5k:
    Type:                    Projected (a volume that contains injected data from multiple sources)
    TokenExpirationSeconds:  3607
    ConfigMapName:           kube-root-ca.crt
    ConfigMapOptional:       <nil>
    DownwardAPI:             true
QoS Class:                   Burstable
Node-Selectors:              kubernetes.io/os=linux
Tolerations:                 CriticalAddonsOnly op=Exists
                             node-role.kubernetes.io/control-plane:NoSchedule
                             node-role.kubernetes.io/master:NoSchedule
                             node.kubernetes.io/not-ready:NoExecute op=Exists for 300s
                             node.kubernetes.io/unreachable:NoExecute op=Exists for 300s
Events:
  Type    Reason    Age    From              Message
  ----    ------    ---    ----              -------
```

Figure 7.26 – When multi-container pods share a network

From the preceding output, we can see CoreDNS using Corefile for configurations. It is located in the following location:

```
/etc/coredns/Corefile
```

We can use the kubectl get configmaps command to inspect the content of Corefile. Here's how it can be done:

```
kubectl get configmaps -n kube-system
```

The output should be as follows:

```
NAME                                      DATA    AGE
coredns                                   1       11h
extension-apiserver-authentication        6       11h
kube-proxy                                2       11h
kube-root-ca.crt                          1       11h
kubeadm-config                            1       11h
kubelet-config                            1       11h
```

Figure 7.27 – When multi-container pods share a network

The preceding command shows there is `configmap` named `coredns`, so let's use the `kubectl describe configmap` command to check out its content:

```
k describe configmap coredns -n kube-system
```

The following output will show how `Corefile` looks:

```
Name:         coredns
Namespace:    kube-system
Labels:       <none>
Annotations:  <none>

Data
====
Corefile:
----
.:53 {
    errors
    health {
        lameduck 5s
    }
    ready
    kubernetes cluster.local in-addr.arpa ip6.arpa {
        pods insecure
        fallthrough in-addr.arpa ip6.arpa
        ttl 30
    }
    prometheus :9153
    forward . /etc/resolv.conf {
        max_concurrent 1000
    }
    cache 30
    loop
    reload
    loadbalance
}


BinaryData
====

Events:  <none>
```

Figure 7.28 – Corefile for CoreDNS

`Corefile` is very useful when you need to customize the DNS resolution process in your Kubernetes cluster. Check out the official documentation about customizing the DNS service here: `https://kubernetes.io/docs/tasks/administer-cluster/dns-custom-nameservers/#coredns-configmap-options`.

Note that the Kubernetes DNS service is registered to the `kubelet` agent, so the Pods running on the cluster use the DNS server's IP address to resolve the DNS names. `kubelet` sets the `/etc/resolv.conf` file for each pod – a DNS query for a `myapp` pod from the `my-packt-apps` namespace can be resolved using either `myapp.my-packt-apps` or `myapp.my-packt-apps.svc.cluster.local`. Now, let's take a closer look at how the DNS hostname works for a pod in a Kubernetes cluster.

Pod IPs and DNS hostnames

Kubernetes creates DNS records for pods. You can contact a pod with fully qualified, consistent DNS hostnames instead of its IP address. For a pod in Kubernetes, the DNS name follows this pattern:

```
<your-pod-ip-address>.<namespace-name>.pod.cluster.local
```

Let's deploy a pod named `nginx` using the following command:

```
kubectl run nginx --image=nginx --port=8080
```

We'll see that the pod has been deployed successfully if you have an output similar to the following:

NAME	READY	STATUS	RESTARTS	AGE
nginx	1/1	Running	0	3s

Let's take a closer look at this pod:

```
kubectl get pod nginx -o wide
```

The output should look as follows:

NAME	READY	STATUS	RESTARTS	AGE	IP	NODE	NOMINATED NODE	READINESS GATES
nginx	1/1	Running	0	2m52s	10.1.0.9	docker-desktop	<none>	<none>

Figure 7.29 – When a multi-container pod shares a network

From the figure, we know the IP address for the `nginx` pod is `10.1.0.9` within the cluster. From the preceding pattern, we could assume that the DNS name of this pod would look as follows:

```
10-1-0-9.default.pod.cluster.local
```

> **Important Note**
>
> Note that in practice, each pod in a StatefulSet derives the hostname from the StatefulSet name. The name domain managed by this service follows this pattern:
>
> ```
> $(service name).$(namespace).svc.cluster.local
> ```
>
> Check out the official documentation to know more: https://kubernetes.io/docs/concepts/workloads/controllers/statefulset/#stable-network-id.

Alternatively, in order to get the IP address of the nginx pod, you can use the kubectl describe pod nginx command, which will open the live detailed spec of your nginx pod. The section called IP is where you can find the pod's IP, as in the following figure:

```
Name:          nginx
Namespace:     default
Priority:      0
Node:          docker-desktop/192.168.65.4
Start Time:    Sat, 11 Jun 2022 20:04:55 -0700
Labels:        run=nginx
Annotations:   <none>
Status:        Running
IP:            10.1.0.9
IPs:
  IP:  10.1.0.9
Containers:
  nginx:
    Container ID:   docker://95993c4e42bb7bcd4c0d37b1b2a8c596fba0331f8006fb73afd2d5d716a0bfdc
    Image:          nginx
    Image ID:       docker-pullable://nginx@sha256:2bcabc23b45489fb0885d69a06ba1d648aeda973fae7bb981bafbb884165e514
    Port:           <none>
    Host Port:      <none>
    State:          Running
      Started:      Sat, 11 Jun 2022 20:04:58 -0700
    Ready:          True
    Restart Count:  0
    Environment:    <none>
    Mounts:
      /var/run/secrets/kubernetes.io/serviceaccount from kube-api-access-zrrvn (ro)
```

Figure 7.30 – When multi-container pods share a network

You can deploy a pod named busybox with the latest Busybox container image in the default namespace and then execute the nslookup command to check out the DNS address of the nginx pod, as shown in the following:

```
kubectl run -it busybox --image=busybox:latest
kubect exec busybox -- nslookup 10.1.0.9
```

The output should look as follows:

```
Server:        10.96.0.10
Address:       10.96.0.10:53

9.0.1.10.in-addr.arpa    name = 10-1-0-9.nginx-svc.default.svc.cluster.local
```

Figure 7.31 – When multi-container pods share a network

Alternatively, you can also use the following command to achieve the same outcome. Note that we are adding two rm flags in the command, which will make sure the pod is deleted once we exit the shell. We also use -- to execute the nslookup command directly. In this way, it allows us to do a quick test, which comes in very handy in the actual CKA exam. The command would look as follows:

```
kubectl run -it sandbox --image=busybox:latest --rm
--restart=Never -- nslookup 10.1.0.9
```

The output should look as follows:

```
Server:        10.96.0.10
Address:       10.96.0.10:53

9.0.1.10.in-addr.arpa    name = 10-1-0-9.nginx-svc.default.svc.cluster.local

pod "sandbox" deleted
```

Figure 7.32 – When multi-container pods share a network

We notice that the only difference is that we get the pod "sandbox" deleted message, which indicates a pod named sandbox gets deleted once we exit the shell. The preceding output shows the DNS name of the nginx pod with the IP address 10.96.0. 10. The PTR record returns the DNS name of this pod as 10-1-0-9.default.pod.cluster.local just as we expected.

Now, let's get the A record of the nginx pod in the default namespace by using the following command:

```
kubectl run -it sandbox --image=busybox:latest --rm
--restart=Never -- nslookup 10-1-0-9.default.pod.cluster.local
```

The output is as follows:

```
Server:    10.96.0.10
Address 1: 10.96.0.10 kube-dns.kube-system.svc.cluster.local

Name:      10-1-0-9.default.pod.cluster.local
Address 1: 10.1.0.9
pod "sandbox" deleted
```

The preceding output proves that the DNS server returns the A record of the nginx pod. Let's deploy a new nginx pod called test-nginx to test out the connectivity by using the following command:

```
$ kubectl run -it test-nginx --image=nginx --rm --restart=Never
-- curl -Is 10-1-0-9.default.pod.cluster.local
```

The output will look as follows:

```
HTTP/1.1 200 OK
Server: nginx/1.21.6
Date: Sun, 12 Jun 2022 18:41:31 GMT
Content-Type: text/html
Content-Length: 615
Last-Modified: Tue, 25 Jan 2022 15:03:52 GMT
Connection: keep-alive
ETag: "61f01158-267"
Accept-Ranges: bytes

pod "test-nginx" deleted
```

Figure 7.33 – When multi-container pods share a network

The preceding screenshot with 200 responses proves that the connectivity between the `test-nginx` pod and `nginx` pod is good and we managed to use the `curl` command on the main page of `nginx` with the DNS name of the `nginx` pod.

Up until this point, we have done a thorough run-through of how IP addresses and DNS work for the pods in a Kubernetes cluster. As we mentioned earlier in this chapter, Kubernetes creates DNS records not only for pods but also for services. Now, let's take a look at how the service IP and DNS work in Kubernetes in the next section.

Service IPs and DNS hostnames

The DNS service in Kubernetes creates DNS records for services so you can contact services with consistent fully qualified DNS hostnames instead of IP addresses. Similarly, for a service in Kubernetes, the DNS follows the following pattern:

```
<service-name>.<namespace-name>.svc.cluster.local
```

Knowing that the DNS server is located in the `kube-system` namespace, we can check it out by using the following command:

```
kubectl get svc -n kube-system
```

The output is as follows, where we can get a look at the IP address of the DNS server in Kubernetes:

```
NAME        TYPE        CLUSTER-IP    EXTERNAL-IP   PORT(S)                  AGE
kube-dns    ClusterIP   10.96.0.10    <none>        53/UDP,53/TCP,9153/TCP   11h
```

Figure 7.34 – When multi-container pods share a network

The preceding screenshot shows the IP address of the DNS server is `10.96.0.10`. Now, let's check out whether we can get the DNS name of the current DNS server by using the following command:

```
kubectl run -it sandbox --image=busybox:latest --rm
--restart=Never -- nslookup 10.96.0.10
```

The output should be as follows:

```
Server:         10.96.0.10
Address:        10.96.0.10:53

10.0.96.10.in-addr.arpa name = kube-dns.kube-system.svc.cluster.local

pod "sandbox" deleted
```

Figure 7.35 – When multi-container pods share a network

The preceding screenshot proves that the DNS name for the DNS server follows the aforementioned pattern from this section. Here is how it looks:

```
kube-dns.kube-system.svc.cluster.local
```

Let's now take a look at exposing a service for the `nginx` pod. We're using the following command to expose the `ClusterIP` service of the `nginx` pod on port `80`:

```
kubectl expose pod nginx --name=nginx-svc --port 80
```

The following output shows that it has been exposed successfully:

```
service/nginx-svc exposed
```

Based on the previous experiment with the `kube-dns` service DNS name, we can expect the `nginx-svc` service to follow the general service DNS name pattern, which will look as follows:

```
nginx-svc.default.svc.cluster.local
```

Now, let's take a look at the services currently in the `default` namespace of our Kubernetes cluster by using the following command:

```
kubectl get svc
```

We can see an output similar to the following:

```
NAME         TYPE        CLUSTER-IP     EXTERNAL-IP   PORT(S)    AGE
kubernetes   ClusterIP   10.96.0.1      <none>        443/TCP    10h
nginx-svc    ClusterIP   10.107.75.83   <none>        80/TCP     59m
```

Figure 7.36 – The services in the Kubernetes default namespace

From the preceding output, we can get a closer look at `nginx-svc` by using the `kubectl get svc nginx-svc -o wide` command. The output is as follows:

```
NAME           TYPE          CLUSTER-IP       EXTERNAL-IP    PORT(S)
AGE      SELECTOR
nginx-svc      ClusterIP     10.107.75.83     <none>         80/TCP
59m    run=nginx
```

The preceding command shows that the IP address of `nginx-svc` is `10.107.75.83`, so let's use the `nslookup` command to check out its DNS name. Use the following command:

```
kubectl run -it sandbox --image=busybox:latest --rm
--restart=Never -- nslookup 10.107.75.83
```

The preceding command will give you the following output:

```
Server:         10.96.0.10
Address:        10.96.0.10:53

83.75.107.10.in-addr.arpa        name = nginx-svc.default.svc.cluster.local

pod "sandbox" deleted
```

Figure 7.37 – Returning the DNS name for nginx-svc by looking up the IP address

Based on the preceding output, we can see that the DNS name for `nginx-svc` is `nginx-svc.default.svc.cluster.local`, which proves our assumption. Let's get the DNS A record of `nginx-service` from the default namespace using the following command:

```
kubectl run -it sandbox --image=busybox:latest --rm
--restart=Never -- nslookup nginx-svc.default.svc.cluster.local
```

You'll see the output is similar to the following:

```
Server:      10.96.0.10
Address 1:   10.96.0.10 kube-dns.kube-system.svc.cluster.local

Name:        nginx-svc.default.svc.cluster.local
Address 1:   10.107.75.83 nginx-svc.default.svc.cluster.local
pod "sandbox" deleted
```

The preceding output shows the DNS server, which was what we saw earlier in this section – the `kube-dns` service with the IP address `10.96.0.10` and under the `kube-dns.kube-system.svc.cluster.local` DNS name. Also, for our `nginx-svc`, we get an IP address of `10.107.75.83` in return.

Now, similar to how we tested the nginx pod, let's test out the connectivity of the nginx service. We can use a pod called challenge-nginx and then run the curl command to see what's coming back. The complete command is as follows:

```
kubectl run -it challenge-nginx --image=nginx --rm
--restart=Never -- curl -Is http://nginx-svc.default.svc.
cluster.local
```

The preceding command leads to the following output:

```
HTTP/1.1 200 OK
Server: nginx/1.21.6
Date: Sun, 12 Jun 2022 19:36:11 GMT
Content-Type: text/html
Content-Length: 615
Last-Modified: Tue, 25 Jan 2022 15:03:52 GMT
Connection: keep-alive
ETag: "61f01158-267"
Accept-Ranges: bytes

pod "challenge-nginx" deleted
```

Figure 7.38 – Returning the DNS name for nginx-svc by looking up the IP address

The preceding screenshot with 200 responses proves the connectivity between the nginx-challenge pod and the nginx-svc service is good, and we managed to use the curl command on the main page of nginx with the DNS name of the nginx service. Knowing the nginx service is exposed from a nginx pod, in real life, we could deploy a number of replicas of this nginx pod, and expose them with one service. The traffic is distributed through the service to each pod.

Summary

This chapter covered Kubernetes networking. It covered the Kubernetes networking model and core networking concepts, as well as how to choose CNI plugins. Working with the Ingress controller and configuring and leveraging CoreDNS in Kubernetes helps you understand how to manage cluster networking and controller access to the applications in Kubernetes.

Make sure you have practiced these examples as you will encounter them often. Notice that this chapter covers 20% of the CKA exam content. Practicing the kubectl commands will help you with better time management, which leads to a greater chance of success in the CKA exam. Together with what we'll talk about in the next chapter about monitoring and logging Kubernetes clusters and applications, you will get a better idea of how to manage Kubernetes clusters in your daily job as a Kubernetes administrator. Stay tuned!

Mock CKA scenario-based practice test

You have two virtual machines, master-0 and worker-0; please complete the following mock scenarios.

Scenario 1

Deploy a new deployment, nginx, with the latest image of nginx for two replicas in a namespace called packt-app. The container is exposed on port 80. Create a service type of ClusterIP within the same namespace. Deploy a sandbox-nginx pod and make a call using curl to verify the connectivity to the nginx service.

Scenario 2

Expose the nginx deployment with the NodePort service type; the container is exposed on port 80. Use the test-nginx pod to make a call using curl to verify the connectivity to the nginx service.

Scenario 3

Make a call using wget or curl from the machine within the same network as that node, to verify the connectivity with the nginx NodePort service through the correct port.

Scenario 4

Use the sandbox-nginx pod and nslookup for the IP address of the nginx NodePort service. See what is returned.

Scenario 5

Use the sandbox-nginx pod and nslookup for the DNS domain hostname of the nginx NodePort service. See what is returned.

Scenario 6

Use the sandbox-nginx pod and nslookup for the DNS domain hostname of the nginx pod. See what is returned.

You can find all the scenario resolutions in *Appendix - Mock CKA scenario-based practice test resolutions* of this book.

FAQs

- *Where can I find the latest updates about Kubernetes networking while working with Kubernetes?*

  The Kubernetes networking **Special Interest Group** (**SIG**) has a GitHub repository that you can follow here: `https://github.com/kubernetes/community/blob/master/sig-network/README.md`.

- *What is the recommended official Kubernetes article for Kubernetes networking?*

  I recommend bookmarking the official documentation about the following topics:

  - Network policy: `https://kubernetes.io/docs/concepts/services-networking/service/`

  - Ingress: `https://kubernetes.io/docs/concepts/services-networking/ingress/`

Part 3:
Troubleshooting

This part covers Kubernetes troubleshooting-related topics ranging from cluster- and application-level logging and monitoring to cluster components and application troubleshooting, security, and networking troubleshooting. This part covers about 30% of the CKA exam's content.

This part of the book comprises the following chapters:

- *Chapter 8, Monitoring and Logging Kubernetes Clusters and Applications*
- *Chapter 9, Troubleshooting Cluster Components and Applications*
- *Chapter 10, Troubleshooting Security and Networking*

8
Monitoring and Logging Kubernetes Clusters and Applications

This chapter describes how to monitor Kubernetes cluster components and applications and get infrastructure-level, system-level, and application-level logs to serve as a source for log analytics or further troubleshooting. Together with the next two chapters about troubleshooting cluster components and applications and troubleshooting Kubernetes security and networking, it covers 30% of the CKA exam content.

In this chapter, we're going to cover the following topics:

- Monitoring on a cluster node
- Monitoring applications on a Kubernetes cluster
- Managing logs at the cluster node and pod levels
- Managing container `stdout` and `stderr` logs

Technical requirements

To get started, you need to make sure your local machine meets the following technical requirements:

- A compatible Linux host. We recommend a Debian-based Linux distribution such as Ubuntu 18.04 or later.
- Make sure your host machine has at least 2 GB of RAM, 2 CPU cores, and about 20 GB of free disk space.

Monitoring on a cluster node

Monitoring is essential for Kubernetes administrators when it comes to getting a clear understanding of what's going on in your Kubernetes cluster. You need to know all of the different metrics to help you get on track in terms of the health of your Kubernetes cluster components. You also need to make sure that your components are operating as expected and that all workloads that are deployed on your worker nodes are functional and have enough resources, such as CPU, memory, and storage. Moreover, you should also check whether any worker nodes are available and have sufficient resources to scale or schedule more workloads.

In Kubernetes, Metrics Server collects CPU/memory metrics and to some extent adjusts the resources needed by containers automatically. Metrics Server collects those metrics every 15 seconds from the kubelet agent and then exposes them in the API server of the Kubernetes master via the Metrics API. This process is described in the following figure:

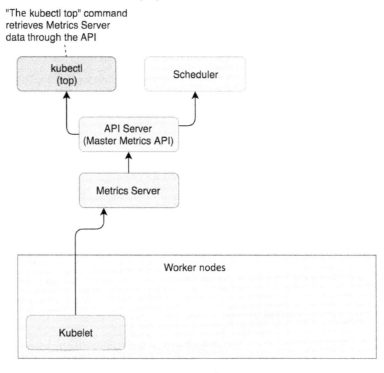

Figure 8.1 – How Metrics Server works in a Kubernetes cluster

Users can use the kubectl top command to access metrics collected by Metrics Server. At the time of writing this chapter, Metrics Server supports scaling up to 5,000 Kubernetes worker nodes, which is the maximum number of nodes that Kubernetes currently supports (Kubernetes v1.24 supports clusters with up to 5,000 nodes). For more details about large Kubernetes clusters, check out this official article: https://kubernetes.io/docs/setup/best-practices/cluster-large/.

Checking whether Metrics Server is installed

From your Kubernetes cluster, you can take the following steps to check whether you have Metrics Server available in your current cluster. You can start by setting up an alias for kubectl using the `alias k=kubectl` command and then use the `k get` command, as follows, to check out the worker nodes that are currently available:

```
alias k=kubectl
k get nodes
```

The preceding command will show the available worker nodes of your current cluster. The output is similar to the following:

```
NAME        STATUS     ROLES                   AGE    VERSION
minikube    Ready      control-plane,master    5d     v1.23.3
```

You can use the `k top node` command to check the metrics for the worker node called `minikube`, as follows:

```
k top node minikube
```

The output of the preceding command will show the resource usage of the `minikube` node if you have Metrics Server installed. Alternatively, you will see the following, which only appears when Metrics Server is not available in your current Kubernetes cluster, which means you need to install Metrics Server:

```
error: Metrics API not available
```

Alternatively, you can use the following command directly to see whether there will be any output:

```
kubectl get pods -n kube-system | grep metrics-server
```

The CKA exam will usually have Metrics Server pre-installed, so you could jump to *step 3* to check out the use cases for the `kubectl top` command.

Installing Metrics Server in your current Kubernetes cluster

If you're on a vanilla Kubernetes cluster, you can install Metrics Server by deploying a YAML definition or through Helm charts; the latter will require Helm to be installed. To get the latest release and instructions, you can go to their GitHub repo: `https://github.com/kubernetes-sigs/metrics-server`.

Using a YAML manifest file

You can use the `kubectl apply -f` command to deploy Metrics Server using the official YAML manifest file as follows:

```
kubectl apply -f https://github.com/kubernetes-sigs/metrics-
server/releases/latest/download/components.yaml
```

Starting from the end of February 2022, there's also a **high-availability** (**HA**) version that bumps up the replica count from one to two for Metrics Server. If you're on a cluster with at least two nodes, you can use the following file:

```
kubectl apply -f https://github.com/kubernetes-sigs/metrics-
server/releases/latest/download/high-availability.yaml
```

You can get more information about Metrics Server here: `https://github.com/kubernetes-sigs/metrics-server/releases`

Using Helm charts

To install Metrics Server using Helm charts, you can go to Artifact Hub and then find the Metrics Server Helm charts at `https://artifacthub.io/packages/helm/metrics-server/metrics-server`

Since Helm 3 is widely used nowadays, you will need to add the Metrics Server Helm charts repo to Helm:

```
helm repo add metrics-server https://kubernetes-sigs.github.io/
metrics-server/
```

It will show the following to confirm that the repo has been added successfully:

```
"metrics-server" has been added to your repositories
```

After adding the repo, you can install the Helm charts through the following command:

```
helm upgrade --install metrics-server metrics-server/metrics-
server
```

The output of the preceding command will show you whether it's been installed successfully.

Using minikube add-ons

If you're using a minikube cluster, Metrics Server comes in the form of a built-in add-on that can be enabled and disabled via the `minikube addons` command. You can use the following to list the currently supported add-ons:

```
minikube addons list
```

The output is similar to what is shown in the following screenshot:

```
cloudmelon@cloudmelonplayground:~$ minikube addons list
|-----------------------------|-----------|------------|-----------------------------------|
|         ADDON NAME          |  PROFILE  |   STATUS   |            MAINTAINER             |
|-----------------------------|-----------|------------|-----------------------------------|
| ambassador                  | minikube  | disabled   | third-party (ambassador)          |
| auto-pause                  | minikube  | disabled   | google                            |
| csi-hostpath-driver         | minikube  | disabled   | kubernetes                        |
| dashboard                   | minikube  | disabled   | kubernetes                        |
| default-storageclass        | minikube  | enabled ✓  | kubernetes                        |
| efk                         | minikube  | disabled   | third-party (elastic)             |
| freshpod                    | minikube  | disabled   | google                            |
| gcp-auth                    | minikube  | disabled   | google                            |
| gvisor                      | minikube  | disabled   | google                            |
| helm-tiller                 | minikube  | disabled   | third-party (helm)                |
| ingress                     | minikube  | disabled   | unknown (third-party)             |
| ingress-dns                 | minikube  | disabled   | google                            |
| istio                       | minikube  | disabled   | third-party (istio)               |
| istio-provisioner           | minikube  | disabled   | third-party (istio)               |
| kong                        | minikube  | disabled   | third-party (Kong HQ)             |
| kubevirt                    | minikube  | disabled   | third-party (kubevirt)            |
| logviewer                   | minikube  | disabled   | unknown (third-party)             |
| metallb                     | minikube  | disabled   | third-party (metallb)             |
| metrics-server              | minikube  | disabled   | kubernetes                        |
| nvidia-driver-installer     | minikube  | disabled   | google                            |
| nvidia-gpu-device-plugin    | minikube  | disabled   | third-party (nvidia)              |
| olm                         | minikube  | disabled   | third-party (operator             |
|                             |           |            | framework)                        |
| pod-security-policy         | minikube  | disabled   | unknown (third-party)             |
| portainer                   | minikube  | disabled   | portainer.io                      |
| registry                    | minikube  | disabled   | google                            |
| registry-aliases            | minikube  | disabled   | unknown (third-party)             |
| registry-creds              | minikube  | disabled   | third-party (upmc enterprises)    |
| storage-provisioner         | minikube  | enabled ✓  | google                            |
| storage-provisioner-gluster | minikube  | disabled   | unknown (third-party)             |
| volumesnapshots             | minikube  | disabled   | kubernetes                        |
|-----------------------------|-----------|------------|-----------------------------------|
```

Figure 8.2 – minikube add-ons list

From the preceding screenshot, we can see the `metrics-server` add-on is `disabled`. You canalso use the following command to get a clearer view:

```
minikube addons list | grep metrics-server
```

The following output shows that currently, the minikube add-on is `disabled`:

```
| metrics-server                        | minikube | disabled        |
kubernetes
```

You can use the `minikube addon enable` command to enable Metrics Server:

```
minikube addons enable metrics-server
```

The following output shows that the Metrics Server add-on was successfully enabled:

```
▪ Using image k8s.gcr.io/metrics-server/metrics-
server:v0.4.2
🌟 The 'metrics-server' addon is enabled
```

Now if you use the `kubectl get` command, you'll see that the Pods and Services related to Metrics Server are up and running in the `kube-system` namespace:

```
kubectl get pod,svc -n kube-system
```

The output should look like the following:

```
[cloudmelon@cloudmelonplayground:~$ kubectl get pod,svc -n kube-system
NAME                                     READY   STATUS    RESTARTS     AGE
pod/coredns-64897985d-brqfl              1/1     Running   0            56d
pod/etcd-minikube                        1/1     Running   0            56d
pod/kube-apiserver-minikube              1/1     Running   0            56d
pod/kube-controller-manager-minikube     1/1     Running   0            56d
pod/kube-proxy-6r287                     1/1     Running   0            56d
pod/kube-scheduler-minikube              1/1     Running   0            56d
pod/metrics-server-6b76bd68b6-rwlb9      1/1     Running   0            78s
pod/storage-provisioner                  1/1     Running   1 (56d ago)  56d

NAME                     TYPE        CLUSTER-IP     EXTERNAL-IP   PORT(S)                  AGE
service/kube-dns         ClusterIP   10.96.0.10     <none>        53/UDP,53/TCP,9153/TCP   56d
service/metrics-server   ClusterIP   10.102.43.189  <none>        443/TCP                  78s
```

Figure 8.3 – Metrics Server Pods and Services in the kube-system namespace

Another command you can use is the following:

```
kubectl get pods -n kube-system | grep metrics-server
```

The output should look like the following:

```
metrics-server-6b76bd68b6-rwlb9     1/1      Running   0
17h
```

As you can see from the output, the Metrics Server pod is up and running, which means you can now use the `kubectl top` command. Let's now take a look at what it does.

Checking out CPU/memory metrics

You can use the `kubectl top` command to top the worker node that you want to get metrics details from. The following is an example where we top a worker node called `minikube`:

```
k top node minikube
```

The output is as follows, where we can see the number of CPU cores and the amount of memory used:

```
NAME        CPU(cores)    CPU%    MEMORY(bytes)    MEMORY%
minikube 232m 11% 961Mi 24%
```

This also applies to the use case where your Kubernetes cluster has multiple worker nodes. Using the `kubectl top node <node name>` command will help you see the resource usage of that specific node.

Monitoring applications on a Kubernetes cluster

A standard end-to-end monitoring solution covers infrastructure monitoring and application monitoring. In Kubernetes, Metrics Server is not only used to monitor the Kubernetes worker nodes but also Kubernetes Pods and containers.

We can test out application monitoring by deploying a new pod in the default namespace as follows:

```
kubectl run nginx --image=nginx
```

After executing the preceding command, make sure that your `nginx` pod is up and running before going to the next section. To check out the status of the pod, you can use the `kubectl get pod nginx` command.

Monitoring the resource usage of an application

You can use the `kubectl top pod <podname>` command to check out the metrics collected for that pod, including the resource consumption of the pod:

```
kubectl top pod nginx
```

The output should look as follows, where you can see the CPU and memory usage of the pod:

```
NAME      CPU(cores)    MEMORY(bytes)
nginx 0m 9Mi
```

In our case, we deployed a single-container pod, but it's important to know that we could also check out the CPU and memory usage for a multi-container pod by using the following command:

```
k top pod < pod name > --containers
```

Let's use the same kubectl top command to show the metrics for the nginx pod and all its containers:

```
k top pod nginx --containers
```

The output should look like the following as it's a single-container pod:

```
POD      NAME     CPU(cores)     MEMORY(bytes)
nginx nginx 0m 9Mi
```

If there are multiple containers, it will list the name of the containers in that pod and show their CPU and memory usage respectively.

With that in mind, we could use kubectl top pod, adding the -A flag or –all-namespaces, to show all the metrics of all the Pods across different namespaces. The following command is used in this case:

```
k top pod -A
```

Alternatively, you can also use the full flag as follows:

```
k top pod --all-namespaces
```

The output should look like the following, where you have all the Pods listed along with their CPU and memory usage respectively:

```
NAMESPACE      NAME                           CPU(cores)
MEMORY(bytes)
default        nginx                          0m
9Mi
kube-system    kube-proxy-64jzv               1m
32Mi
kube-system    kube-proxy-hplp5               1m
28Mi
kube-system    kube-proxy-kvb96               2m
31Mi
kube-system    kube-proxy-kvjwh               1m
28Mi
kube-system    kube-proxy-rmw2r               1m
```

```
31Mi
kube-system      kube-proxy-tcz5m                         1m
26Mi
kube-system      metrics-server-6576d9ccf8-z8mlg          6m
37M
```

There's a good chance that the CKA exam will ask you what pod consumes the most compute resources in a list of pods, or any other task of this nature – that's where the `–sort-by` flag comes into play. The `--sort-by` flag accepts either `cpu` or `memory` as a value, and as a result it will return the result `asc` or `desc`. The command looks as in the following examples:

```
kubectl top pod --sort-by=cpu
kubectl top pod --sort-by=memory
```

It makes more sense when we have a large list of pods and you have requested to sort them by the memory or CPU resources consumed, from most to least. We can use the following command to do this:

```
kubectl top pod -A --sort-by=memory
```

The output should look as follows, with all the pods across all the namespaces in your current Kubernetes cluster listed according to resource usage:

```
kube-system      metrics-server-6576d9ccf8-z8mlg          7m
37Mi
kube-system      kube-proxy-64jzv                         1m
32Mi
kube-system      kube-proxy-rmw2r                         1m
31Mi
kube-system      kube-proxy-kvb96                         1m
31Mi
kube-system      kube-proxy-kvjwh                         1m
28Mi
kube-system      kube-proxy-hplp5                         1m
28Mi
kube-system      kube-proxy-tcz5m                         1m
25Mi
default          nginx                                    0m
9Mi
```

This command works in a similar way when using `–sort-by  cpu` flag. The output lists the pods in the order of most CPU consumed to least.

Checking application details

You can use the `kubectl describe pod <podname>` command to find out status information regarding the allocated CPUs and memory usage and some other information, such as runtime versions, system information, capacity, labels, and annotations:

```
kubectl describe pod nginx
```

The output should look like the following:

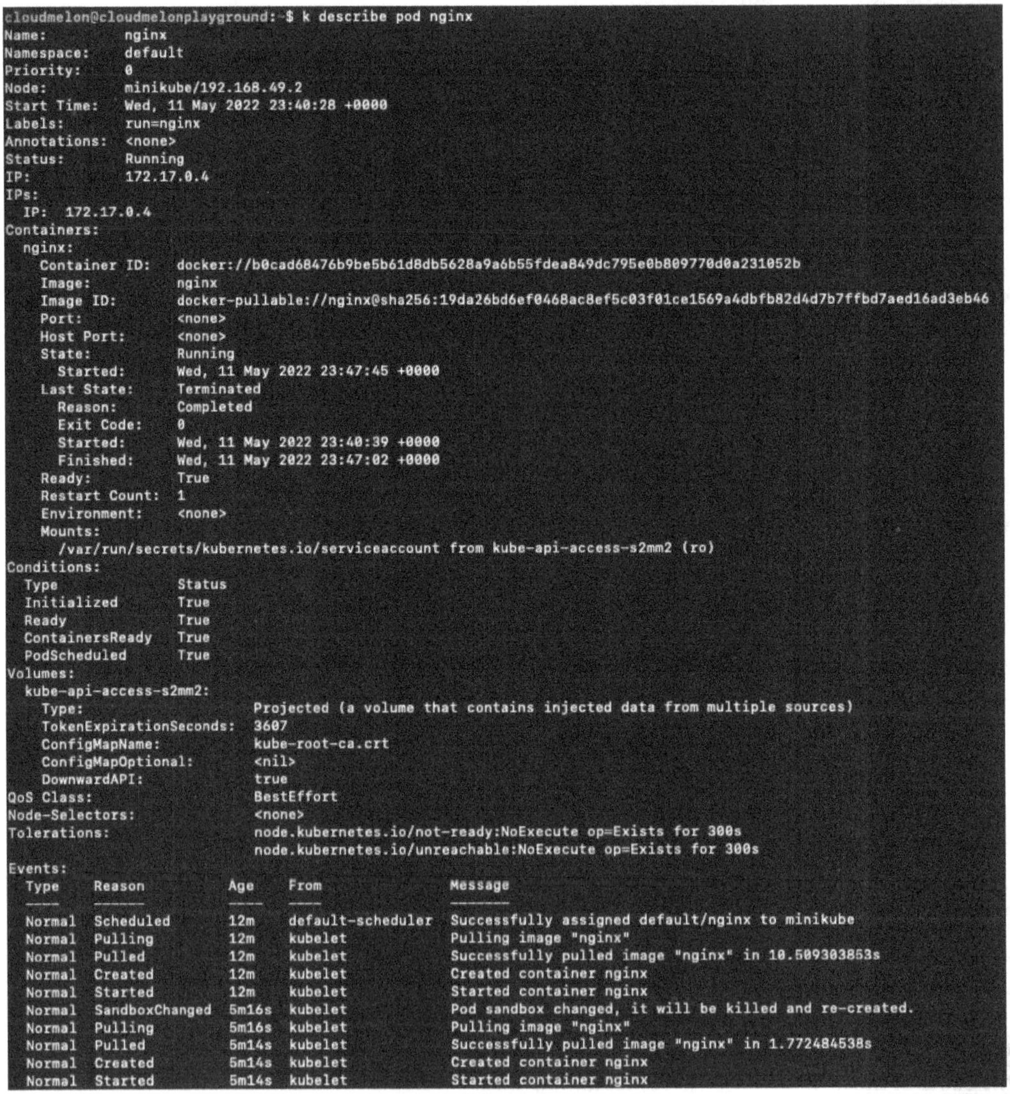

Figure 8.4 – kubectl describe pod nginx

Note that there's an `Events` section at the bottom of the preceding screenshot that shows a log of recent events related to this pod. We'll take a closer look at the `Events` section:

```
Events:
  Type     Reason          Age    From               Message
  ----     ------          ---    ----               -------
  Normal   Scheduled       12m    default-scheduler  Successfully assigned default/nginx to minikube
  Normal   Pulling         12m    kubelet            Pulling image "nginx"
  Normal   Pulled          12m    kubelet            Successfully pulled image "nginx" in 10.509303853s
  Normal   Created         12m    kubelet            Created container nginx
  Normal   Started         12m    kubelet            Started container nginx
  Normal   SandboxChanged  5m16s  kubelet            Pod sandbox changed, it will be killed and re-created.
  Normal   Pulling         5m16s  kubelet            Pulling image "nginx"
  Normal   Pulled          5m14s  kubelet            Successfully pulled image "nginx" in 1.772484538s
  Normal   Created         5m14s  kubelet            Created container nginx
  Normal   Started         5m14s  kubelet            Started container nginx
```

Figure 8.5 – Events of the nginx pod

The events here include a series of events in Kubernetes, such as these:

1. The pod gets scheduled to the worker node called `minikube`.
2. The container image is pulled from the container registry.
3. The kubelet agent provisions the pod containing an `nginx` container.
4. Kubelet starts the pod and the `nginx` container starts to accept traffic.

Analyzing those events helps us to understand what's going on during the pod provisioning process, and it could give us clues as to whether any exceptions happened and why, allowing us to come up with potential solutions. We'll take a closer look at the events in the next section of this chapter.

If a pod is in a namespace other than the default namespace, you can specify the -n flag in the `kubectl` `describe` command to add the namespace. The following is an example using this command to describe a pod named `coredns-64897985d-brqf1` in the `kube-system` namespace:

```
kubectl describe pod coredns-64897985d-brqf1 -n kube-system
```

The output should look like the following:

```
Name:                  coredns-64897985d-brqf1
Namespace:             kube-system
Priority:              2000000000
Priority Class Name:   system-cluster-critical
Node:                  minikube/192.168.49.2
Start Time:            Mon, 14 Mar 2022 00:25:47 +0000
Labels:                k8s-app=kube-dns
                       pod-template-hash=64897985d
Annotations:           <none>
Status:                Running
IP:                    172.17.0.2
IPs:
  IP:          172.17.0.2
Controlled By:  ReplicaSet/coredns-64897985d
Containers:
  coredns:
    Container ID:  docker://c8b57ce2dd8daa5a3c0ba2d282ce3a8c3ed789381bc9fd1e52b95d72338fb277
    Image:         k8s.gcr.io/coredns/coredns:v1.8.6
    Image ID:      docker-pullable://k8s.gcr.io/coredns/coredns@sha256:5b6ec0d6de9baaf3e92d0f66cd96a25b9edbce8716f5f15dcd1a616b3abd590e
    Ports:         53/UDP, 53/TCP, 9153/TCP
    Host Ports:    0/UDP, 0/TCP, 0/TCP
    Args:
      -conf
      /etc/coredns/Corefile
    State:          Running
      Started:      Mon, 14 Mar 2022 00:25:48 +0000
    Ready:          True
    Restart Count:  0
    Limits:
      memory:  170Mi
    Requests:
      cpu:     100m
      memory:  70Mi
    Liveness:   http-get http://:8080/health delay=60s timeout=5s period=10s #success=1 #failure=5
    Readiness:  http-get http://:8181/ready delay=0s timeout=1s period=10s #success=1 #failure=3
    Environment:  <none>
    Mounts:
      /etc/coredns from config-volume (ro)
      /var/run/secrets/kubernetes.io/serviceaccount from kube-api-access-jn4vd (ro)
Conditions:
  Type              Status
  Initialized       True
  Ready             True
  ContainersReady   True
  PodScheduled      True
Volumes:
  config-volume:
    Type:      ConfigMap (a volume populated by a ConfigMap)
    Name:      coredns
    Optional:  false
  kube-api-access-jn4vd:
    Type:                    Projected (a volume that contains injected data from multiple sources)
    TokenExpirationSeconds:  3607
    ConfigMapName:           kube-root-ca.crt
    ConfigMapOptional:       <nil>
    DownwardAPI:             true
QoS Class:       Burstable
Node-Selectors:  kubernetes.io/os=linux
Tolerations:     CriticalAddonsOnly op=Exists
                 node-role.kubernetes.io/control-plane:NoSchedule
                 node-role.kubernetes.io/master:NoSchedule
                 node.kubernetes.io/not-ready:NoExecute op=Exists for 300s
                 node.kubernetes.io/unreachable:NoExecute op=Exists for 300s
Events:          <none>
```

Figure 8.6 – kubectl describe coredns pod in the kube-system namespace

Even though the preceding screenshots contain similar chunks of information, the details differ from pod to pod. You could add `> mypod.yaml` to the end of the command to export the pod information for further analysis:

```
kubectl describe pod nginx > mypod.yaml
```

You will get a YAML file called `mypod.yaml` containing critical pod information.

Monitoring cluster events

We can get Kubernetes events by using the following command:

```
kubectl get events
```

We can get events logged in the current cluster, which includes events logged previously in the Events section, when we use the `kubectl describe pod` command. The following is a sample output after running the `kubectl get events` command:

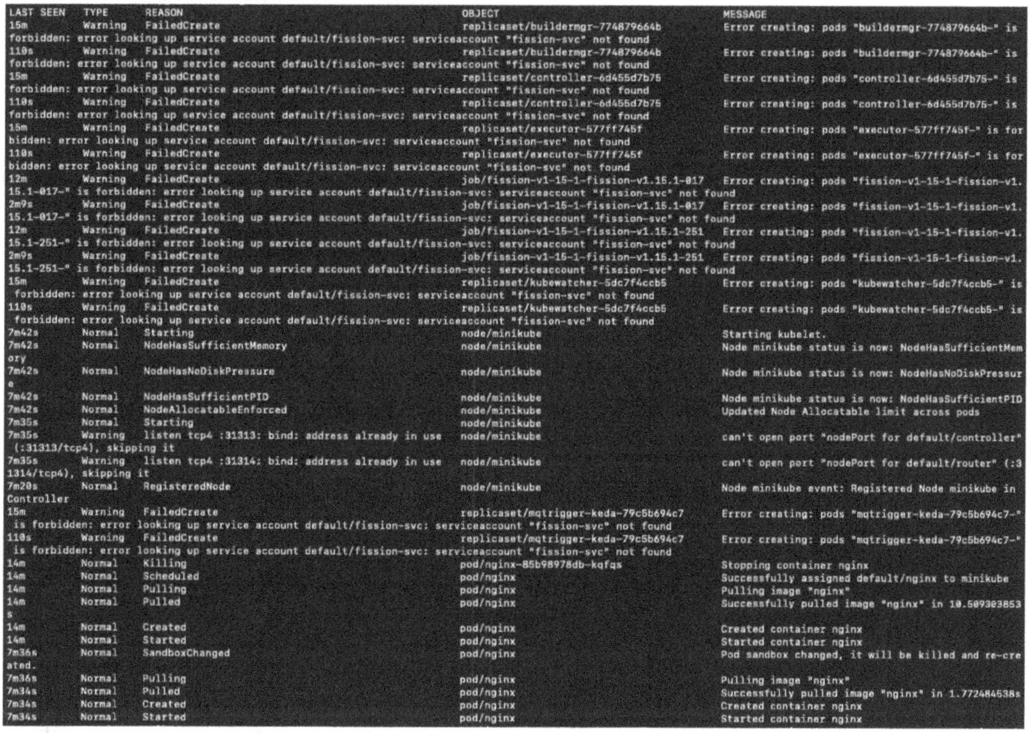

Figure 8.7 – kubectl get events

You can use the following command to list the events sorted by timestamp:

```
kubectl get events --sort-by=.metadata.creationTimestamp
```

If you want to collect the events during a deployment, you can run the following command on the side:

```
kubectl get events --watch
```

The commands will give you a good idea of what's going on during the deployment process if you're not using Kubernetes Dashboard or any third-party monitoring frameworks such as Prometheus with Grafana Dashboard. Knowing about what happens at the application level by monitoring sometimes comes in handy, especially when it comes to troubleshooting. Often we get a better understanding by analyzing logs and tracking exceptions. Let's take a look at how to manage logs at the cluster node and pod levels.

Managing logs at the cluster node and Pod levels

Logs are very handy when it comes to troubleshooting issues. The information collected in a log is usually helpful in understanding what has happened, figuring out why certain issues happened, and finding remediations to prevent them from happening again later on.

Cluster-level logging

In Kubernetes, the notion of cluster-level logging is widely recognized. This means logs are meant to be stored in a separate backend, so the lifecycles of those logs are independent of what's been logged down to the worker node, pod, or even container level.

Kubernetes itself does not provide a comprehensive native logging framework, but it can be integrated with lots of third-party open source logging solutions in the community, such as Grafana Loki or the EFK stack, which includes Elasticsearch, Fluentd, and Kibana for log searching, querying, and tracing.

Logging in Kubernetes involves a set of patterns that are implemented by the community with different open source solutions. There are the following three patterns:

- **Using a node-level logging agent that runs on every node**: The agent is often in a DaemonSet so it will be evenly distributed on each node, and this agent pushes the logs to a backend. In this case, there are no code changes for the application.

- **Using a dedicated sidecar container to log information from the application in the same Pod**: This case can be in conjunction with a logging agent running on the node or streaming the logs out, and it is usually recommended to write log entries with the same formats to the same log stream for convenient processing.

- **Directly streaming the logs from the application to an external backend**: This can work with external object storage, as such storage supports lifecycle policies, which allows the setup of data retention policies and the archiving of old logs based on the policy. Most object storage also works with a search framework, where logs are indexed and so are easy to search and query.

To learn more about the Kubernetes logging architecture, check this article out: `https://kubernetes.io/docs/concepts/cluster-administration/logging/`

Checking out the node details

With native Kubernetes, you can use the `kubectl describe node <nodename>` command to find out the status information regarding the allocated CPUs and memory usage as well as some other information, such as runtime versions, system information, capacity, labels, and annotations. We can use the following command to describe a worker node named `minikube`:

```
kubectl describe node minikube
```

The output is similar to the following:

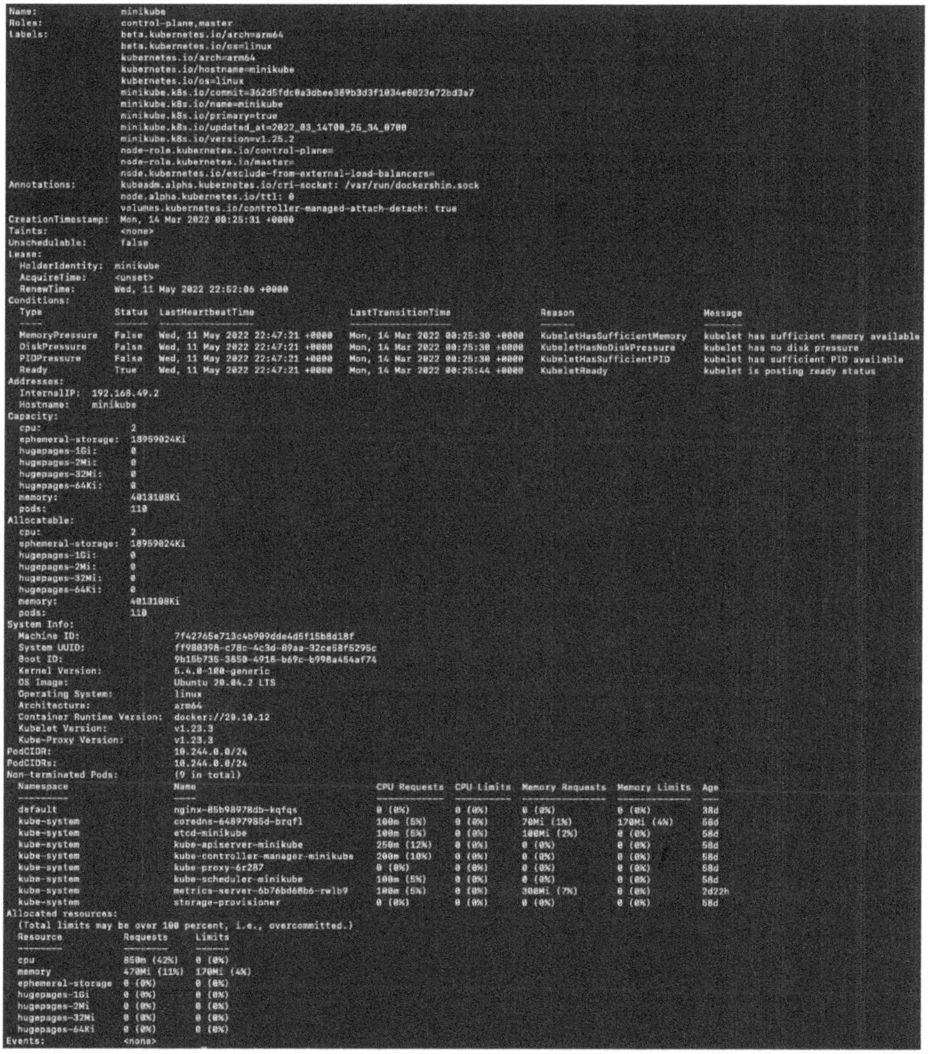

```
Name:               minikube
Roles:              control-plane,master
Labels:             beta.kubernetes.io/arch=arm64
                    beta.kubernetes.io/os=linux
                    kubernetes.io/arch=arm64
                    kubernetes.io/hostname=minikube
                    kubernetes.io/os=linux
                    minikube.k8s.io/commit=362d5fdc0a3dbee389b3d3f1034e0023e72bd3a7
                    minikube.k8s.io/name=minikube
                    minikube.k8s.io/primary=true
                    minikube.k8s.io/updated_at=2022_03_14T00_25_34_0700
                    minikube.k8s.io/version=v1.25.2
                    node-role.kubernetes.io/control-plane=
                    node-role.kubernetes.io/master=
                    node.kubernetes.io/exclude-from-external-load-balancers=
Annotations:        kubeadm.alpha.kubernetes.io/cri-socket: /var/run/dockershim.sock
                    node.alpha.kubernetes.io/ttl: 0
                    volumes.kubernetes.io/controller-managed-attach-detach: true
CreationTimestamp:  Mon, 14 Mar 2022 00:25:31 +0000
Taints:             <none>
Unschedulable:      false
Lease:
  HolderIdentity:  minikube
  AcquireTime:     <unset>
  RenewTime:       Wed, 11 May 2022 22:52:06 +0000
Conditions:
  Type             Status  LastHeartbeatTime                 LastTransitionTime                Reason                       Message
  ----             ------  -----------------                 ------------------                ------                       -------
  MemoryPressure   False   Wed, 11 May 2022 22:47:21 +0000   Mon, 14 Mar 2022 00:25:30 +0000   KubeletHasSufficientMemory   kubelet has sufficient memory available
  DiskPressure     False   Wed, 11 May 2022 22:47:21 +0000   Mon, 14 Mar 2022 00:25:30 +0000   KubeletHasNoDiskPressure     kubelet has no disk pressure
  PIDPressure      False   Wed, 11 May 2022 22:47:21 +0000   Mon, 14 Mar 2022 00:25:30 +0000   KubeletHasSufficientPID      kubelet has sufficient PID available
  Ready            True    Wed, 11 May 2022 22:47:21 +0000   Mon, 14 Mar 2022 00:25:44 +0000   KubeletReady                 kubelet is posting ready status
Addresses:
  InternalIP:  192.168.49.2
  Hostname:    minikube
Capacity:
  cpu:                2
  ephemeral-storage:  18959024Ki
  hugepages-1Gi:      0
  hugepages-2Mi:      0
  hugepages-32Mi:     0
  hugepages-64Ki:     0
  memory:             4013108Ki
  pods:               110
Allocatable:
  cpu:                2
  ephemeral-storage:  18959024Ki
  hugepages-1Gi:      0
  hugepages-2Mi:      0
  hugepages-32Mi:     0
  hugepages-64Ki:     0
  memory:             4013108Ki
  pods:               110
System Info:
  Machine ID:                 7f42765e713c4b909dde4d5f15b8d18f
  System UUID:                ff980398-c78c-4c3d-89aa-32ce58f5295c
  Boot ID:                    9b15b735-3850-4916-b69c-b998a454af74
  Kernel Version:             5.4.0-100-generic
  OS Image:                   Ubuntu 20.04.2 LTS
  Operating System:           linux
  Architecture:               arm64
  Container Runtime Version:  docker://20.10.12
  Kubelet Version:            v1.23.3
  Kube-Proxy Version:         v1.23.3
PodCIDR:                      10.244.0.0/24
PodCIDRs:                     10.244.0.0/24
Non-terminated Pods:          (9 in total)
  Namespace                   Name                                CPU Requests  CPU Limits  Memory Requests  Memory Limits  Age
  ---------                   ----                                ------------  ----------  ---------------  -------------  ---
  default                     nginx-85b98978db-kqfqs              0 (0%)        0 (0%)      0 (0%)           0 (0%)         38d
  kube-system                 coredns-64897985d-brqfl             100m (5%)     0 (0%)      70Mi (1%)        170Mi (4%)     58d
  kube-system                 etcd-minikube                       100m (5%)     0 (0%)      100Mi (2%)       0 (0%)         58d
  kube-system                 kube-apiserver-minikube             250m (12%)    0 (0%)      0 (0%)           0 (0%)         58d
  kube-system                 kube-controller-manager-minikube    200m (10%)    0 (0%)      0 (0%)           0 (0%)         58d
  kube-system                 kube-proxy-6r287                    0 (0%)        0 (0%)      0 (0%)           0 (0%)         58d
  kube-system                 kube-scheduler-minikube             100m (5%)     0 (0%)      0 (0%)           0 (0%)         58d
  kube-system                 metrics-server-6b76bd68b6-rwlb9     100m (5%)     0 (0%)      300Mi (7%)       0 (0%)         2d22h
  kube-system                 storage-provisioner                 0 (0%)        0 (0%)      0 (0%)           0 (0%)         58d
Allocated resources:
  (Total limits may be over 100 percent, i.e., overcommitted.)
  Resource           Requests     Limits
  --------           --------     ------
  cpu                850m (42%)   0 (0%)
  memory             470Mi (11%)  170Mi (4%)
  ephemeral-storage  0 (0%)       0 (0%)
  hugepages-1Gi      0 (0%)       0 (0%)
  hugepages-2Mi      0 (0%)       0 (0%)
  hugepages-32Mi     0 (0%)       0 (0%)
  hugepages-64Ki     0 (0%)       0 (0%)
Events:              <none>
```

Figure 8.8 – kubectl describe node minikube

Gettting to know the node specification will give you an understanding of how your node was previously configured. Let's now take a look at how to get some quick but handy information using the kubectl describe node command.

Checking the node status

With the kubectl describe command, we get some general information about a node. Notice that it also contains an events section that usually logs node events. To get more status information from a node, we usually use the following command, taking a node named minikube as an example:

```
kubectl get node minikube -o wide
```

The output is similar to the following:

Figure 8.9 – kubectl get node output

From the preceding screenshot, if you compare the kubectl get node command with the one with the -o wide flag, you'll see that it gives extra information about the image and kernel version as well as the container runtime, which is quite handy when we need to get information quickly.

Managing container stdout and stderr logs

In the Unix and Linux OSs, there are three I/O streams, called STDIN, STDOUT, and STDERR. Here, we'll talk about STDOUT and STERR in Linux containers, which are typically what the kubectl logs command shows to us.

STDOUT is usually a command's normal output, and STDERR is typically used to output error messages. Kubernetes uses the kubectl logs <podname> command to log STDOUT and STDERR. It looks like the following when we use the command to log the nginx pod that we deployed in this chapter:

```
kubectl logs nginx
```

The output should look like the following:

```
cloudmelon@cloudmelonplayground:~$ k logs nginx
/docker-entrypoint.sh: /docker-entrypoint.d/ is not empty, will attempt to perform configuration
/docker-entrypoint.sh: Looking for shell scripts in /docker-entrypoint.d/
/docker-entrypoint.sh: Launching /docker-entrypoint.d/10-listen-on-ipv6-by-default.sh
10-listen-on-ipv6-by-default.sh: info: Getting the checksum of /etc/nginx/conf.d/default.conf
10-listen-on-ipv6-by-default.sh: info: Enabled listen on IPv6 in /etc/nginx/conf.d/default.conf
/docker-entrypoint.sh: Launching /docker-entrypoint.d/20-envsubst-on-templates.sh
/docker-entrypoint.sh: Launching /docker-entrypoint.d/30-tune-worker-processes.sh
/docker-entrypoint.sh: Configuration complete; ready for start up
2022/05/11 23:47:45 [notice] 1#1: using the "epoll" event method
2022/05/11 23:47:45 [notice] 1#1: nginx/1.21.6
2022/05/11 23:47:45 [notice] 1#1: built by gcc 10.2.1 20210110 (Debian 10.2.1-6)
2022/05/11 23:47:45 [notice] 1#1: OS: Linux 5.4.0-100-generic
2022/05/11 23:47:45 [notice] 1#1: getrlimit(RLIMIT_NOFILE): 1048576:1048576
2022/05/11 23:47:45 [notice] 1#1: start worker processes
2022/05/11 23:47:45 [notice] 1#1: start worker process 32
2022/05/11 23:47:45 [notice] 1#1: start worker process 33
```

Figure 8.10 – kubectl logs nginx pod

Now, we'll use a container to write text to the standard output stream with a frequency of once per second. We can do this by deploying a new pod. The following is an example of a YAML manifest for this pod:

```
apiVersion: v1
kind: Pod
metadata:
  name: logger
spec:
  containers:
  - name: packs
    image: busybox:1.28
    args: [/bin/sh, -c,
            'i=0; while true; do echo "$i: $(date)";
i=$((i+1)); sleep 1; done']
```

You can use the kubectl logs command to retrieve the logs from the logger Pod as follows:

```
k logs logger
```

The log would look as follows:

```
0: Thu May 12 04:34:40 UTC 2022
1: Thu May 12 04:34:41 UTC 2022
2: Thu May 12 04:34:42 UTC 2022
3: Thu May 12 04:34:43 UTC 2022
```

We can get into the pod to retrieve the specific container log by using the `- c` flag. Let's check out the log for a container called `packt` in the `logger` pod using the following command:

```
k logs logger -c packt
```

The following output is the logs retrieved from the `packt` container:

```
cloudmelon@cloudmelonplayground:~$ k logs counter -c count
0: Thu May 12 04:34:40 UTC 2022
1: Thu May 12 04:34:41 UTC 2022
2: Thu May 12 04:34:42 UTC 2022
3: Thu May 12 04:34:43 UTC 2022
4: Thu May 12 04:34:44 UTC 2022
5: Thu May 12 04:34:45 UTC 2022
6: Thu May 12 04:34:46 UTC 2022
7: Thu May 12 04:34:47 UTC 2022
8: Thu May 12 04:34:48 UTC 2022
9: Thu May 12 04:34:49 UTC 2022
10: Thu May 12 04:34:50 UTC 2022
11: Thu May 12 04:34:51 UTC 2022
12: Thu May 12 04:34:52 UTC 2022
13: Thu May 12 04:34:53 UTC 2022
14: Thu May 12 04:34:54 UTC 2022
15: Thu May 12 04:34:55 UTC 2022
16: Thu May 12 04:34:56 UTC 2022
```

Figure 8.11 – Logs from the packt container

If you want to stream the logs, you can use the `kubectl logs -f` command, as follows:

```
kubectl logs -f logger
```

You should be able to see an output like the following:

```
11: Thu May 12 04:34:51 UTC 2022
12: Thu May 12 04:34:52 UTC 2022
13: Thu May 12 04:34:53 UTC 2022
14: Thu May 12 04:34:54 UTC 2022
15: Thu May 12 04:34:55 UTC 2022
16: Thu May 12 04:34:56 UTC 2022
17: Thu May 12 04:34:57 UTC 2022
18: Thu May 12 04:34:58 UTC 2022
19: Thu May 12 04:34:59 UTC 2022
20: Thu May 12 04:35:00 UTC 2022
21: Thu May 12 04:35:01 UTC 2022
22: Thu May 12 04:35:02 UTC 2022
23: Thu May 12 04:35:03 UTC 2022
24: Thu May 12 04:35:04 UTC 2022
25: Thu May 12 04:35:05 UTC 2022
26: Thu May 12 04:35:06 UTC 2022
27: Thu May 12 04:35:07 UTC 2022
28: Thu May 12 04:35:08 UTC 2022
29: Thu May 12 04:35:09 UTC 2022
30: Thu May 12 04:35:10 UTC 2022
31: Thu May 12 04:35:11 UTC 2022
32: Thu May 12 04:35:12 UTC 2022
33: Thu May 12 04:35:13 UTC 2022
34: Thu May 12 04:35:14 UTC 2022
35: Thu May 12 04:35:15 UTC 2022
36: Thu May 12 04:35:16 UTC 2022
37: Thu May 12 04:35:17 UTC 2022
38: Thu May 12 04:35:18 UTC 2022
39: Thu May 12 04:35:19 UTC 2022
40: Thu May 12 04:35:20 UTC 2022
41: Thu May 12 04:35:21 UTC 2022
42: Thu May 12 04:35:22 UTC 2022
43: Thu May 12 04:35:23 UTC 2022
44: Thu May 12 04:35:24 UTC 2022
45: Thu May 12 04:35:25 UTC 2022
46: Thu May 12 04:35:26 UTC 2022
```

Figure 8.12 – kubectl logs for the nginx pod

Use the following command if you want to return logs newer than a certain duration, such as within 1 hour:

```
kubectl logs --since=1h
```

You can modify the value after the -since flag as per your requirements.

Summary

This chapter covered monitoring and logging for Kubernetes on three levels – cluster, node, and pod. This chapter laid the groundwork for the next two chapters, where we will focus on troubleshooting cluster components and application failures, as well as exploring some other challenges around Kubernetes security restrictions and container networking by providing more specific troubleshooting use cases and end-to-end troubleshooting scenarios. Stay tuned!

Mock CKA scenario-based practice test

You have two virtual machines, master-0 and worker-0: please complete the following mock scenarios.

Scenario 1

List all the available Pods in your current cluster, identify the ones with the highest CPU consumption, and write their names to a max-cpu.txt file.

You can find all the scenario resolutions in *Appendix - Mock CKA scenario-based practice test resolutions* of this book.

FAQs

- *Where can I find out about the latest updates on Kubernetes Metrics Server?*

  Kubernetes Metrics Server has a GitHub repository at https://github.com/kubernetes-sigs/metrics-server.

- *Where can I find the latest information on Kubernetes cluster logging architecture?*

  Go to the official Kubernetes documentation at https://kubernetes.io/docs/concepts/cluster-administration/logging/.

- *Where can I find the metrics for Kubernetes system components?*

  You can bookmark this page to get more information: https://kubernetes.io/docs/concepts/cluster-administration/system-metrics/.

9

Troubleshooting Cluster Components and Applications

Troubleshooting is one of the main tasks performed during your daily work as a Kubernetes administrator. This chapter introduces the general approaches to troubleshooting errors caused by cluster component failure and the issues that can occur during application deployments.

In this chapter, we're going to cover the following topics:

- Kubernetes troubleshooting general practices
- Troubleshooting cluster components
- Troubleshooting applications

Technical requirements

To get started, we need to make sure our local machine meets the technical requirements described as follows.

In case you're on Linux, see the following:

- A compatible Linux host. We recommend a Debian-based Linux distribution such as Ubuntu 18.04 or later.
- Make sure your host machine has at least 2 GB RAM, 2 CPU cores, and about 20 GB of free disk space.

In case you're on Windows 10 or Windows 11, see the following:

- We recommend updating Docker Desktop to the latest version and creating a Docker Desktop local Kubernetes cluster. Check out this article to learn about how to set up a local Kubernetes cluster with Docker Desktop: `https://docs.docker.com/desktop/kubernetes/`.

- We also recommend using **Windows Subsystem for Linux 2 (WSL 2)** to test the environment. Refer to this article to see how to install WSL (`https://docs.microsoft.com/en-us/windows/wsl/install`) and the following article to see how to set up the Docker Desktop WSL 2 backend: `https://docs.docker.com/desktop/windows/wsl/`.

Once you're set up, you can check whether you're currently set to the correct Kubernetes cluster using the following command:

```
alias k=kubectlk config current-context
```

The preceding command will print out the current cluster in the output. In our case, it was similar to the following, as we're on Windows with a Kubernetes local cluster created by Docker Desktop:

```
docker desktop
```

If you've been following our demonstration along the way in this book, you'll have noticed that most of the demonstration was on a `minikube` cluster. In this case, the output would be the following:

```
minikube
```

You may have used your local machine to connect with a few different Kubernetes clusters – you can use `kubectl config view` to check which is the current cluster:

```
apiVersion: v1
clusters:
- cluster:
    certificate-authority-data: DATA+OMITTED
    server: https://kubernetes.docker.internal:6443
  name: docker-desktop
contexts:
- context:
    cluster: docker-desktop
    user: docker-desktop
  name: docker-desktop
current-context: docker-desktop
kind: Config
preferences: {}
users:
- name: docker-desktop
  user:
    client-certificate-data: REDACTED
    client-key-data: REDACTED
```

Figure 9.1 – Local cluster context information

To learn more about how to organize cluster access using `kubeconfig` and how to configure access to multiple clusters, refer to *Chapter 6, Securing Kubernetes*.

In this chapter, we will use `docker-desktop` to understand how to troubleshoot local Kubernetes clusters. Note that the same set of commands is also applied to `minikube`. Let's start by talking about the general practice of Kubernetes troubleshooting.

General practices in Kubernetes troubleshooting

We have talked about the common tasks performed as a part of the daily job as a Kubernetes administrator a lot in this book, especially in the previous chapters. In real life, upon the stage of the project that you're involved in, a Kubernetes administrator is likely to be involved in the installation and set-up of Kubernetes cluster phase, applications deployment, and managing the security and networking aspects of things for Kubernetes. In addition to the aforementioned tasks, operating and maintaining Kubernetes clusters and applications deployed on the cluster also form some of the key responsibilities of a Kubernetes administrator. Therefore, acquiring good troubleshooting skills will greatly help in this scenario.

Troubleshooting Kubernetes clusters is a combination of identifying, diagnosing, and remediating an issue – the problem statement covers Kubernetes cluster components, nodes, networking, and security. Additionally, the problem statement also covers the application level, such as pods, or even the container level. We'll cover troubleshooting Kubernetes cluster components and the application level, including pods and containers, in this chapter.

It's important to take an outside-in approach and gradually narrow down the scope to identify the root cause of an issue. This means we can rationalize the process using the following recommendations:

- Monitoring plays a vital role in identifying potential problems and finding their root causes. In *Chapter 8, Monitoring and Logging Kubernetes Clusters and Applications*, we covered how to monitor Kubernetes cluster components, as well as applications, together with the instructions about logging, which helps you make your first steps.

- Metrics analysis is the first step shortly after you detect a potential issue. Although sometimes the problem statements may not be as they seem, you can make the troubleshooting easier by starting with analyzing metrics from the cluster and node level to get a high-level view, then moving down to the application.

- Sometimes, metrics may not tell you the whole story. In this case, analyzing the logs will help you piece the information together better. At this point, if you find that you have a better idea about the issue that occurred, it's about time to dive deep into those logs and find the root cause, as compared to the one you thought was the culprit. However, it's still a good idea to go back one level higher to see whether anything in the process was missing.

- Once you have found the issue, an actionable remediation plan is required if you want to prevent the issue from ever happening again, rather than just applying a quick fix to the issue. This step will contribute to your future success and make your daily job much easier. Maintenance and troubleshooting work becomes a daily operation task after the initial setup – it is a key component of your daily job as a Kubernetes administrator.

In the actual CKA exam, troubleshooting holds more weight and some of the given scenarios are quite time-consuming, as it is usually stressful to find the root cause within a limited time window. However, as a candidate, you can confidently plan your time ahead once you're certain about the fact that you have done an overall great job with the other high-value questions, such as the ones about application deployment, networking, and backup etcd storage. The troubleshooting exam questions usually appear in the second half of the CKA exam – you can usually start by analyzing the Kubernetes cluster components. There is a higher chance the questions will be about `kubelet` on the worker node and then escalate to the application level. Be mindful of performing the troubleshooting and fixing the issue on the correct node before moving on.

Based on the aforementioned outside-in approach, let's talk about troubleshooting the cluster component first.

Troubleshooting cluster components

Troubleshooting cluster components includes the Kubernetes system processes on the master node and worker node. We'll take a look at some common troubleshooting scenarios in this section and will be starting from a higher-level view.

Inspecting the cluster

Inspecting the cluster and node is usually the first step toward detecting the issues on the control plane. We can do that using the following command:

```
kubectl cluster-info
```

The output renders the addresses of the control plane components and services:

```
Kubernetes control plane is running at https://kubernetes.docker.internal:6443
CoreDNS is running at https://kubernetes.docker.internal:6443/api/v1/namespaces/kube-system/services/kube-dns:dns/proxy

To further debug and diagnose cluster problems, use 'kubectl cluster-info dump'.
```

Figure 9.2 – Rendering the cluster information

If you want further information for debugging and diagnosis, use the following command:

```
kubectl cluster-info dump
```

The preceding command gives an output that is huge and contains a lot of information – hence, we've only displayed the key part in the following screenshot:

```
==== START logs for container echoserver of pod default/echoserver ====
failed to try resolving symlinks in path "/var/log/pods/default_echoserver_a6c40e19-2b04-48ab-b948-ed4c085fe0e0/echoserver/0.log": ls
tat /var/log/pods/default_echoserver_a6c40e19-2b04-48ab-b948-ed4c085fe0e0: no such file or directory==== END logs for container echos
erver of pod default/echoserver ====
==== START logs for container nginx of pod default/nginx ====
/docker-entrypoint.sh: /docker-entrypoint.d/ is not empty, will attempt to perform configuration
/docker-entrypoint.sh: Looking for shell scripts in /docker-entrypoint.d/
/docker-entrypoint.sh: Launching /docker-entrypoint.d/10-listen-on-ipv6-by-default.sh
10-listen-on-ipv6-by-default.sh: info: Getting the checksum of /etc/nginx/conf.d/default.conf
10-listen-on-ipv6-by-default.sh: info: Enabled listen on IPv6 in /etc/nginx/conf.d/default.conf
/docker-entrypoint.sh: Launching /docker-entrypoint.d/20-envsubst-on-templates.sh
/docker-entrypoint.sh: Launching /docker-entrypoint.d/30-tune-worker-processes.sh
/docker-entrypoint.sh: Configuration complete; ready for start up
2022/06/19 04:05:53 [notice] 1#1: using the "epoll" event method
2022/06/19 04:05:53 [notice] 1#1: nginx/1.21.6
2022/06/19 04:05:53 [notice] 1#1: built by gcc 10.2.1 20210110 (Debian 10.2.1-6)
2022/06/19 04:05:53 [notice] 1#1: OS: Linux 5.10.102.1-microsoft-standard-WSL2
2022/06/19 04:05:53 [notice] 1#1: getrlimit(RLIMIT_NOFILE): 1048576:1048576
2022/06/19 04:05:53 [notice] 1#1: start worker processes
2022/06/19 04:05:53 [notice] 1#1: start worker process 31
2022/06/19 04:05:53 [notice] 1#1: start worker process 32
2022/06/19 04:05:53 [notice] 1#1: start worker process 33
2022/06/19 04:05:53 [notice] 1#1: start worker process 34
2022/06/19 04:05:53 [notice] 1#1: start worker process 35
2022/06/19 04:05:53 [notice] 1#1: start worker process 36
2022/06/19 04:05:53 [notice] 1#1: start worker process 37
2022/06/19 04:05:53 [notice] 1#1: start worker process 38
2022/06/19 04:05:53 [notice] 1#1: start worker process 39
2022/06/19 04:05:53 [notice] 1#1: start worker process 40
2022/06/19 04:05:53 [notice] 1#1: start worker process 41
2022/06/19 04:05:53 [notice] 1#1: start worker process 42
2022/06/19 04:05:53 [notice] 1#1: start worker process 43
2022/06/19 04:05:53 [notice] 1#1: start worker process 44
2022/06/19 04:05:53 [notice] 1#1: start worker process 45
2022/06/19 04:05:53 [notice] 1#1: start worker process 46
2022/06/19 04:05:53 [notice] 1#1: start worker process 47
2022/06/19 04:05:53 [notice] 1#1: start worker process 48
2022/06/19 04:05:53 [notice] 1#1: start worker process 49
2022/06/19 04:05:53 [notice] 1#1: start worker process 50
==== END logs for container nginx of pod default/nginx ====
```

Figure 9.3 – The Kubernetes cluster logs

The preceding screenshot shows the log information and is very helpful for finding the root causes. Although we could get good information out of the control plane and cluster logs, you'll get errors for the workloads running on top of it quite often, which can happen because of the node availability or capability. Let's take a look at troubleshooting approaches with the node in the next section.

Inspecting the node

Inspecting the node using the following command will help you get the current state of your current cluster and nodes:

```
kubectl get nodes
```

The output should look as follows:

```
NAME             STATUS   ROLES           AGE     VERSION
docker-desktop   Ready    control-plane   7d10h   v1.24.0
```

Figure 9.4 – The Kubernetes node information

The preceding screenshot shows that the only worker node that we have here is in the `Ready` status. When you have multiple nodes, you will see a list of nodes in the output.

The `ROLES` column shows the role of your node – it could be a `control-plane`, `etcd`, or `worker`:

- The `control-plane` role runs the Kubernetes master components, besides `etcd`.
- The `etcd` role runs the etcd store. Refer to *Chapter 3, Maintaining Kubernetes Clusters*, to learn more about the etcd store.
- The `worker` role runs the Kubernetes worker node – that's where your containerized workloads land.

The `STATUS` column shows the current condition of the running nodes – the ideal status that we all love is `Ready`. Examples of the possible conditions are listed in the following table:

Node condition	What does that mean?
`Ready`	The node is healthy and ready to accept pods.
`DiskPressure`	The disk capacity is low.
`MemoryPressure`	The node memory is low.
`PIDPressure`	Too many processes are running on the node.
`NetworkUnavailable`	The networking is incorrectly configured.
`SchedulingDisabled`	This is not a condition in the Kubernetes API but it appears after you cordon a node. Refer to *Chapter 3, Maintaining Kubernetes Clusters*, to learn about how to perform a version upgrade on a Kubernetes cluster using `kubeadm` when you need to cordon the nodes.

Table 9.1 - Different node conditions and what they mean

Another column that is very interesting from the aforementioned output is the `VERSION` column – this one shows the Kubernetes version running on this node. Kubernetes versions here mean the Kubernetes master components version, the etcd version, or `kubelet` version, vary from node role to node role. Refer to *Chapter 3, Maintaining Kubernetes Clusters*, to learn about upgrading versions on the Kubernetes nodes.

In case you do have suspicions about the node, you can use the following command to inspect the node information:

```
kubectl describe node docker-desktop
```

The output should be similar to the following. As you can see, you can get more detailed information from this as compared to the `kubectl get node` command:

```
Name:                docker-desktop
Roles:               control-plane
Labels:              beta.kubernetes.io/arch=amd64
                     beta.kubernetes.io/os=linux
                     kubernetes.io/arch=amd64
                     kubernetes.io/hostname=docker-desktop
                     kubernetes.io/os=linux
                     node-role.kubernetes.io/control-plane=
                     node.kubernetes.io/exclude-from-external-load-balancers=
Annotations:         kubeadm.alpha.kubernetes.io/cri-socket: unix:///var/run/cri-dockerd.sock
                     node.alpha.kubernetes.io/ttl: 0
                     volumes.kubernetes.io/controller-managed-attach-detach: true
CreationTimestamp:   Sat, 11 Jun 2022 10:18:45 -0700
Taints:              <none>
Unschedulable:       false
Lease:
  HolderIdentity:    docker-desktop
  AcquireTime:       <unset>
  RenewTime:         Sat, 18 Jun 2022 21:10:51 -0700
Conditions:
  Type             Status  LastHeartbeatTime                 LastTransitionTime                Reason                       Message
  ----             ------  -----------------                 ------------------                ------                       -------
  MemoryPressure   False   Sat, 18 Jun 2022 21:10:52 -0700   Sat, 11 Jun 2022 10:18:43 -0700   KubeletHasSufficientMemory   kubelet has sufficient memory available
  DiskPressure     False   Sat, 18 Jun 2022 21:10:52 -0700   Sat, 11 Jun 2022 10:18:43 -0700   KubeletHasNoDiskPressure     kubelet has no disk pressure
  PIDPressure      False   Sat, 18 Jun 2022 21:10:52 -0700   Sat, 11 Jun 2022 10:18:43 -0700   KubeletHasSufficientPID      kubelet has sufficient PID available
  Ready            True    Sat, 18 Jun 2022 21:10:52 -0700   Sat, 11 Jun 2022 10:19:16 -0700   KubeletReady                 kubelet is posting ready status
Addresses:
  InternalIP:  192.168.65.4
  Hostname:    docker-desktop
Capacity:
  cpu:                20
  ephemeral-storage:  263174212Ki
  hugepages-1Gi:      0
  hugepages-2Mi:      0
  memory:             52382988Ki
  pods:               110
Allocatable:
  cpu:                20
  ephemeral-storage:  242541353378
  hugepages-1Gi:      0
  hugepages-2Mi:      0
  memory:             52280588Ki
  pods:               110
System Info:
  Machine ID:                 f572738f-ad78-4ffe-b9ba-e2d406999e3e
  System UUID:                f572738f-ad78-4ffe-b9ba-e2d406999e3e
  Boot ID:                    29ef0153-f41b-434b-86af-7be2f1260d8c
  Kernel Version:             5.10.102.1-microsoft-standard-WSL2
  OS Image:                   Docker Desktop
  Operating System:           linux
  Architecture:               amd64
  Container Runtime Version:  docker://20.10.16
  Kubelet Version:            v1.24.0
  Kube-Proxy Version:         v1.24.0
Non-terminated Pods:          (17 in total)
  Namespace      Name                                    CPU Requests  CPU Limits  Memory Requests  Memory Limits  Age
  ---------      ----                                    ------------  ----------  ---------------  -------------  ---
                 default                 nginx                         0 (0%)      0 (0%)           0 (0%)         0 (0%)          7d1h
  default        nginx-8f458dc5b-p74rr                   0 (0%)        0 (0%)      0 (0%)           0 (0%)         6d6h
  default        nginx-8f458dc5b-v8j74                   0 (0%)        0 (0%)      0 (0%)           0 (0%)         6d6h
  default        nginx-v2                                0 (0%)        0 (0%)      0 (0%)           0 (0%)         7d1h
  default        test                                    0 (0%)        0 (0%)      0 (0%)           0 (0%)         7d
  default        webfront-app-5c474b5bcc-nsvtt           0 (0%)        0 (0%)      0 (0%)           0 (0%)         6d4h
  default        webfront-app-5c474b5bcc-zrwq2           0 (0%)        0 (0%)      0 (0%)           0 (0%)         6d4h
  kube-system    coredns-6d4b75cb6d-4h89j                100m (0%)     0 (0%)      70Mi (0%)        170Mi (0%)     6d23h
  kube-system    coredns-6d4b75cb6d-kj6cq                100m (0%)     0 (0%)      70Mi (0%)        170Mi (0%)     7d18h
  kube-system    etcd-docker-desktop                     100m (0%)     0 (0%)      100Mi (0%)       0 (0%)         7d18h
  kube-system    kube-apiserver-docker-desktop           250m (1%)     0 (0%)      0 (0%)           0 (0%)         7d18h
```

Figure 9.5 – The kubectl describe node output information

To get the most value out of the preceding command, we could check out the `Conditions` section, which should look as follows:

```
Conditions:
  Type             Status  LastHeartbeatTime                 LastTransitionTime                Reason                       Message
  ----             ------  -----------------                 ------------------                ------                       -------
  MemoryPressure   False   Tue, 21 Jun 2022 15:13:26 -0700   Sat, 11 Jun 2022 10:18:43 -0700   KubeletHasSufficientMemory   kubelet has sufficient memory available
  DiskPressure     False   Tue, 21 Jun 2022 15:13:26 -0700   Sat, 11 Jun 2022 10:18:43 -0700   KubeletHasNoDiskPressure     kubelet has no disk pressure
  PIDPressure      False   Tue, 21 Jun 2022 15:13:26 -0700   Sat, 11 Jun 2022 10:18:43 -0700   KubeletHasSufficientPID      kubelet has sufficient PID available
  Ready            True    Tue, 21 Jun 2022 15:13:26 -0700   Sat, 11 Jun 2022 10:19:16 -0700   KubeletReady                 kubelet is posting ready status
```

Figure 9.6 – Getting the node condition information

The preceding screenshots show the detailed node condition information, as we explained earlier in this chapter. It is also possible to get the allocated resource information from the same output, which shows the following:

```
Allocated resources:
  (Total limits may be over 100 percent, i.e., overcommitted.)
  Resource           Requests       Limits
  --------           --------       ------
  cpu                950m (4%)      100m (0%)
  memory             290Mi (0%)     390Mi (0%)
  ephemeral-storage  0 (0%)         0 (0%)
  hugepages-1Gi      0 (0%)         0 (0%)
  hugepages-2Mi      0 (0%)         0 (0%)
```

Figure 9.7 – Getting the node resource consumption information

The value from the preceding screenshot is to understand the current consumption of the cluster in terms of CPU, memory, and storage.

The same output also helps you get an overview of the resource requests and limits from the individual pods running in the current cluster, as shown in the following screenshot:

Namespace	Name	CPU Requests	CPU Limits	Memory Requests	Memory Limits	Age
Non-terminated Pods:	(13 in total)					
default	nginx	0 (0%)	0 (0%)	0 (0%)	0 (0%)	9d
default	nginx-v2	0 (0%)	0 (0%)	0 (0%)	0 (0%)	9d
default	test	0 (0%)	0 (0%)	0 (0%)	0 (0%)	9d
kube-system	coredns-6d4b75cb6d-4h89j	100m (0%)	0 (0%)	70Mi (0%)	170Mi (0%)	9d
kube-system	coredns-6d4b75cb6d-kj6cq	100m (0%)	0 (0%)	70Mi (0%)	170Mi (0%)	10d
kube-system	etcd-docker-desktop	100m (0%)	0 (0%)	100Mi (0%)	0 (0%)	10d
kube-system	kube-apiserver-docker-desktop	250m (1%)	0 (0%)	0 (0%)	0 (0%)	10d
kube-system	kube-controller-manager-docker-desktop	200m (1%)	0 (0%)	0 (0%)	0 (0%)	10d
kube-system	kube-flannel-ds-vnh4k	100m (0%)	100m (0%)	50Mi (0%)	50Mi (0%)	8d
kube-system	kube-proxy-9rfxs	0 (0%)	0 (0%)	0 (0%)	0 (0%)	10d
kube-system	kube-scheduler-docker-desktop	100m (0%)	0 (0%)	0 (0%)	0 (0%)	10d
kube-system	storage-provisioner	0 (0%)	0 (0%)	0 (0%)	0 (0%)	10d
kube-system	vpnkit-controller	0 (0%)	0 (0%)	0 (0%)	0 (0%)	10d

Figure 9.8 – Get the pod resource consumption information

If you want to envision this output in a more structured way, you can use the following command to make it look more similar to a yaml file:

```
kubectl get node docker-desktop -o yaml
```

The output is as follows:

```yaml
apiVersion: v1
kind: Node
metadata:
  annotations:
    kubeadm.alpha.kubernetes.io/cri-socket: unix:///var/run/cri-dockerd.sock
    node.alpha.kubernetes.io/ttl: "0"
    volumes.kubernetes.io/controller-managed-attach-detach: "true"
  creationTimestamp: "2022-06-11T17:18:45Z"
  labels:
    beta.kubernetes.io/arch: amd64
    beta.kubernetes.io/os: linux
    kubernetes.io/arch: amd64
    kubernetes.io/hostname: docker-desktop
    kubernetes.io/os: linux
    node-role.kubernetes.io/control-plane: ""
    node.kubernetes.io/exclude-from-external-load-balancers: ""
  name: docker-desktop
  resourceVersion: "465114"
  uid: d9a38207-0c3c-4ba7-bf58-1ceed2ffdd3c
spec: {}
status:
  addresses:
  - address: 192.168.65.4
    type: InternalIP
  - address: docker-desktop
    type: Hostname
  allocatable:
    cpu: "20"
    ephemeral-storage: "242541353378"
    hugepages-1Gi: "0"
    hugepages-2Mi: "0"
    memory: 52280588Ki
    pods: "110"
  capacity:
    cpu: "20"
    ephemeral-storage: 263174212Ki
    hugepages-1Gi: "0"
    hugepages-2Mi: "0"
    memory: 52382988Ki
```

Figure 9.9 – Getting the node information in YAML

With the preceding output, pay attention in particular to the section called `nodeInfo`, which gives you an overview of the OS image, architecture, kernel version, `kubeProxy` version, `kubelet` version, and os:

```
nodeInfo:
  architecture: amd64
  bootID: 29ef0153-f41b-434b-86af-7be2f1260d8c
  containerRuntimeVersion: docker://20.10.16
  kernelVersion: 5.10.102.1-microsoft-standard-WSL2
  kubeProxyVersion: v1.24.0
  kubeletVersion: v1.24.0
  machineID: f572738f-ad78-4ffe-b9ba-e2d406999e3e
  operatingSystem: linux
  osImage: Docker Desktop
  systemUUID: f572738f-ad78-4ffe-b9ba-e2d406999e3e
```

Figure 9.10 – Getting pod resource consumption information

In case you don't want that full overview of the Kubernetes node and want to focus on getting the memory of the current running process in your Kubernetes cluster, you can run the following command within the Kubernetes node:

```
top
```

The output is refined and should look similar to the following:

```
top - 17:53:47 up 20:48,  0 users,  load average: 0.20, 0.24, 0.25
Tasks:  10 total,   1 running,   9 sleeping,   0 stopped,   0 zombie
%Cpu(s):  1.3 us,  0.6 sy,  0.0 ni, 97.9 id,  0.0 wa,  0.0 hi,  0.1 si,  0.0 st
MiB Mem :  51155.3 total,  46595.1 free,   2129.5 used,   2430.6 buff/cache
MiB Swap:  13312.0 total,  13312.0 free,      0.0 used.  48497.8 avail Mem

   PID USER       PR  NI    VIRT    RES    SHR S  %CPU  %MEM     TIME+ COMMAND
     8 root       20   0     896     88     20 S   0.3   0.0   0:00.22 init
  1133 cloudme+   20   0   10872   3752   3204 R   0.3   0.0   0:00.02 top
     1 root       20   0     904    540    464 S   0.0   0.0   0:00.01 init
     7 root       20   0     896     88     20 S   0.0   0.0   0:00.00 init
     9 cloudme+   20   0   10056   5068   3364 S   0.0   0.0   0:00.18 bash
   524 root       20   0     904     96     20 S   0.0   0.0   0:00.00 init
   525 root       20   0     904     96     20 S   0.0   0.0   0:00.00 init
   526 root       20   0 1976028  30976  13600 S   0.0   0.1   0:02.12 docker-desktop-
   543 root       20   0     904     96     20 S   0.0   0.0   0:00.00 init
   544 cloudme+   20   0  766320  49652  30084 S   0.0   0.1   0:04.22 docker
```

Figure 9.11 – Checking on the consumption information of the processes

As we explained earlier in this chapter, `DiskPressure` is also a key factor in the health status of the worker node. You can use the following command to check the available disk storage:

```
df -h
```

The output looks similar to the following:

```
Filesystem      Size  Used Avail Use% Mounted on
/dev/sdc        251G  2.2G  237G   1% /
tmpfs            25G     0   25G   0% /mnt/wsl
tools           953G  319G  635G  34% /init
none             25G     0   25G   0% /dev
none             25G   12K   25G   1% /run
none             25G     0   25G   0% /run/lock
none             25G     0   25G   0% /run/shm
none             25G     0   25G   0% /run/user
tmpfs            25G     0   25G   0% /sys/fs/cgroup
drivers         953G  319G  635G  34% /usr/lib/wsl/drivers
lib             953G  319G  635G  34% /usr/lib/wsl/lib
C:\             953G  319G  635G  34% /mnt/c
/dev/sdd        251G  3.4G  235G   2% /mnt/wsl/docker-desktop-data/isocache
none             25G   16K   25G   1% /mnt/wsl/docker-desktop/shared-sockets/host-services
/dev/sdb        251G  121M  239G   1% /mnt/wsl/docker-desktop/docker-desktop-user-distro
/dev/loop0      350M  350M     0 100% /mnt/wsl/docker-desktop/cli-tools
```

Figure 9.12 – The available disk information

After checking on the cluster and node information, we can go to the next step, which is checking on the Kubernetes components.

Inspecting the Kubernetes components

We could make this checking easier and more effective by examining the processes in the kube-system namespace – that's where you'll find most of them and be able to export some handy information such as configurations, diagnosis logs, and so on.

Troubleshooting a system-reserved process

Check for errors in a system-reserved process using the following command:

```
kubectl get pods -n kube-system
```

In case you have multiple nodes, you can add the `-o wide` flag to see which pods are running on which node:

```
kubectl get pods -n kube-system -o wide
```

As you may already know from the previous chapters, this command will print out the system-reserved processes:

```
NAME                                        READY   STATUS    RESTARTS       AGE
coredns-6d4b75cb6d-4h89j                    1/1     Running   2 (2d4h ago)   9d
coredns-6d4b75cb6d-kj6cq                    1/1     Running   2 (2d4h ago)   10d
etcd-docker-desktop                         1/1     Running   2 (2d4h ago)   10d
kube-apiserver-docker-desktop               1/1     Running   2 (2d4h ago)   10d
kube-controller-manager-docker-desktop      1/1     Running   2 (2d4h ago)   10d
kube-flannel-ds-p7rhw                       1/1     Running   0              5s
kube-proxy-9rfxs                            1/1     Running   2 (2d4h ago)   10d
kube-scheduler-docker-desktop               1/1     Running   2 (2d4h ago)   10d
storage-provisioner                         1/1     Running   4 (2d4h ago)   10d
```

Figure 9.13 – The system-reserved process

When you see any process that is not in the Running status, it means that it was unhealthy – you can use the kubectl describe pod command to check on it. The following is an example to check out the kube-proxy status:

```
k describe pod kube-proxy-9rfxs -n kube-system
```

The preceding command will print out the full descriptive information of the kube-proxy-9rfxs pod. However, as this pod presents the kube-proxy component, we can narrow the pod information down further by using the following command:

```
k describe pod kube-proxy-9rfxs -n kube-system | grep Node:
```

The output prints out the node name and its allocated IP address:

```
Node:                   docker-desktop/192.168.65.4
```

You can double-check this by using the kubectl get node -o wide command, which will print out the IP address of the docker-desktop node too. It provides the same IP address as the following (here is a partial output):

```
NAME             STATUS   ROLES           AGE   VERSION   INTERNAL-IP
docker-desktop   Ready    control-plane   10d   v1.24.0   192.168.65.4
```

Figure 9.14 – Node-related information

From the output of the kubectl describe pod, kube-proxy-9rfxs -n kubectl, we know the kube-proxy is a DaemonSet – refer to *Chapter 4, Application Scheduling and Lifecycle Management*, to refresh the details about DaemonSets. In the case that you have multiple nodes and want to see which pod is on which node, you can also use the following command to check out your kube-proxy DaemonSet:

```
kubectl describe daemonset kube-proxy -n kube-system
```

The output is similar to the following, in which you can find useful information such as `Pod Status` and `pod template`, which shows you the details of this pod:

```
Name:              kube-proxy
Selector:          k8s-app=kube-proxy
Node-Selector:     kubernetes.io/os=linux
Labels:            k8s-app=kube-proxy
Annotations:       deprecated.daemonset.template.generation: 1
Desired Number of Nodes Scheduled: 1
Current Number of Nodes Scheduled: 1
Number of Nodes Scheduled with Up-to-date Pods: 1
Number of Nodes Scheduled with Available Pods: 1
Number of Nodes Misscheduled: 0
Pods Status:  1 Running / 0 Waiting / 0 Succeeded / 0 Failed
Pod Template:
  Labels:            k8s-app=kube-proxy
  Service Account:   kube-proxy
  Containers:
   kube-proxy:
    Image:       k8s.gcr.io/kube-proxy:v1.24.0
    Port:        <none>
    Host Port:   <none>
    Command:
      /usr/local/bin/kube-proxy
      --config=/var/lib/kube-proxy/config.conf
      --hostname-override=$(NODE_NAME)
    Environment:
      NODE_NAME:   (v1:spec.nodeName)
    Mounts:
      /lib/modules from lib-modules (ro)
      /run/xtables.lock from xtables-lock (rw)
      /var/lib/kube-proxy from kube-proxy (rw)
  Volumes:
   kube-proxy:
    Type:      ConfigMap (a volume populated by a ConfigMap)
    Name:      kube-proxy
    Optional:  false
   xtables-lock:
    Type:         HostPath (bare host directory volume)
    Path:         /run/xtables.lock
    HostPathType: FileOrCreate
   lib-modules:
    Type:         HostPath (bare host directory volume)
    Path:         /lib/modules
    HostPathType:
  Priority Class Name:  system-node-critical
Events:                 <none>
```

Figure 9.15 – The kube-proxy DaemonSet information

Knowing the pod configuration from the preceding output is not enough. When the pod is not up and running for some reason, the logs are much handier, especially when the Events section is none (as can be seen in the preceding screenshot). We can use the following command to check the pod logs:

```
kubectl logs kube-proxy-9rfxs -n kube-system
```

The preceding command prints out logs similar to the following, which will give you more details about what has happened:

```
E0619 19:12:32.983236       1 proxier.go:657] "Failed to read builtin modules file, you can ignore this message when kube-proxy is ru
nning inside container without mounting /lib/modules" err="open /lib/modules/5.10.102.1-microsoft-standard-WSL2/modules.builtin: no s
uch file or directory" filePath="/lib/modules/5.10.102.1-microsoft-standard-WSL2/modules.builtin"
I0619 19:12:32.984728       1 proxier.go:667] "Failed to load kernel module with modprobe, you can ignore this message when kube-prox
y is running inside container without mounting /lib/modules" moduleName="ip_vs"
I0619 19:12:32.985952       1 proxier.go:667] "Failed to load kernel module with modprobe, you can ignore this message when kube-prox
y is running inside container without mounting /lib/modules" moduleName="ip_vs_rr"
I0619 19:12:32.987044       1 proxier.go:667] "Failed to load kernel module with modprobe, you can ignore this message when kube-prox
y is running inside container without mounting /lib/modules" moduleName="ip_vs_wrr"
I0619 19:12:32.988348       1 proxier.go:667] "Failed to load kernel module with modprobe, you can ignore this message when kube-prox
y is running inside container without mounting /lib/modules" moduleName="ip_vs_sh"
I0619 19:12:32.989398       1 proxier.go:667] "Failed to load kernel module with modprobe, you can ignore this message when kube-prox
y is running inside container without mounting /lib/modules" moduleName="nf_conntrack"
I0619 19:12:32.997581       1 node.go:163] Successfully retrieved node IP: 192.168.65.4
I0619 19:12:32.997625       1 server_others.go:138] "Detected node IP" address="192.168.65.4"
I0619 19:12:32.997657       1 server_others.go:578] "Unknown proxy mode, assuming iptables proxy" proxyMode=""
I0619 19:12:33.017568       1 server_others.go:206] "Using iptables Proxier"
I0619 19:12:33.017598       1 server_others.go:213] "kube-proxy running in dual-stack mode" ipFamily=IPv4
I0619 19:12:33.017604       1 server_others.go:214] "Creating dualStackProxier for iptables"
I0619 19:12:33.017608       1 server_others.go:485] "Detect-local-mode set to ClusterCIDR, but no cluster CIDR defined"
I0619 19:12:33.017612       1 server_others.go:541] "Defaulting to no-op detect-local" detect-local-mode="ClusterCIDR"
I0619 19:12:33.017628       1 proxier.go:259] "Setting route_localnet=1, use nodePortAddresses to filter loopback addresses for NodeP
orts to skip it https://issues.k8s.io/90259"
I0619 19:12:33.017763       1 proxier.go:259] "Setting route_localnet=1, use nodePortAddresses to filter loopback addresses for NodeP
orts to skip it https://issues.k8s.io/90259"
I0619 19:12:33.017941       1 server.go:661] "Version info" version="v1.24.0"
I0619 19:12:33.017960       1 server.go:663] "Golang settings" GOGC="" GOMAXPROCS="" GOTRACEBACK=""
I0619 19:12:33.018221       1 conntrack.go:100] "Set sysctl" entry="net/netfilter/nf_conntrack_tcp_timeout_close_wait" value=3600
I0619 19:12:33.018474       1 config.go:444] "Starting node config controller"
I0619 19:12:33.018515       1 shared_informer.go:255] Waiting for caches to sync for node config
I0619 19:12:33.018529       1 config.go:317] "Starting service config controller"
I0619 19:12:33.018543       1 shared_informer.go:255] Waiting for caches to sync for service config
I0619 19:12:33.018658       1 config.go:226] "Starting endpoint slice config controller"
I0619 19:12:33.018686       1 shared_informer.go:255] Waiting for caches to sync for endpoint slice config
I0619 19:12:33.118902       1 shared_informer.go:262] Caches are synced for endpoint slice config
I0619 19:12:33.118943       1 shared_informer.go:262] Caches are synced for service config
I0619 19:12:33.118911       1 shared_informer.go:262] Caches are synced for node config
```

Figure 9.16 – The pod logs information

After covering master node troubleshooting, when troubleshooting is needed in the worker node, we should start by troubleshooting the kubelet agent – let's get into this in the next section.

Troubleshooting the kubelet agent

After checking on the node status, we could SSH to that worker node if you're not there already, and use the following command to check on the kubelet status:

```
systemctl status kubelet
```

The output should look as follows:

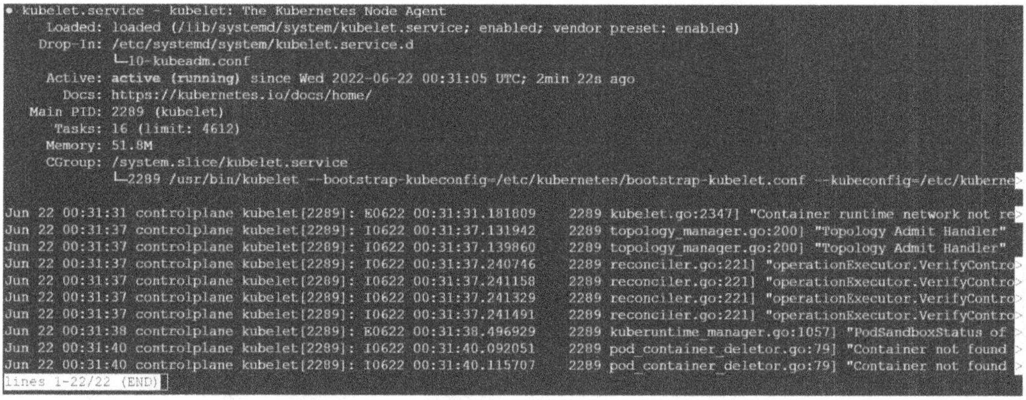

Figure 9.17 – The kubelet agent status and logs

The important part of the preceding screenshot is the status of kubelet, as can be seen in the following screenshot:

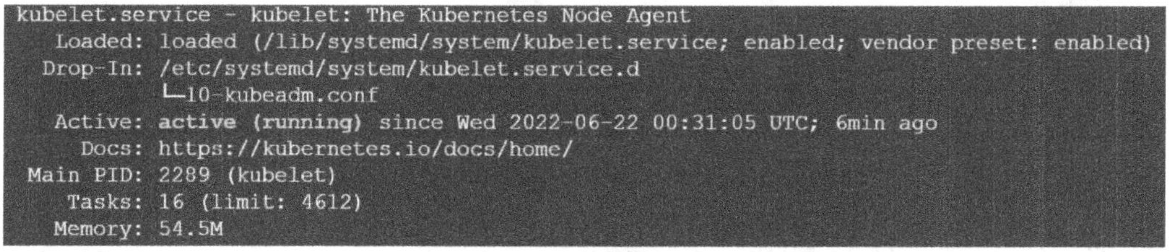

Figure 9.18 – The kubelet agent status

In the case that the status is not active (running), we could use journalctl to obtain the logs on the kubelet service on the worker node. The following command shows how to do so:

```
journalctl -u kubelet.service
```

The output will print out log details similar to the following:

```
-- Logs begin at Wed 2022-06-22 00:30:33 UTC, end at Wed 2022-06-22 00:35:11 UTC. --
Jun 22 00:30:37 host01 systemd[1]: Started kubelet: The Kubernetes Node Agent.
Jun 22 00:30:39 controlplane kubelet[863]: E0622 00:30:39.592428     863 server.go:205] "Failed to load kubelet config file">
Jun 22 00:30:39 controlplane systemd[1]: kubelet.service: Main process exited, code=exited, status=1/FAILURE
Jun 22 00:30:39 controlplane systemd[1]: kubelet.service: Failed with result 'exit-code'.
Jun 22 00:30:41 controlplane systemd[1]: Stopped kubelet: The Kubernetes Node Agent.
Jun 22 00:30:50 controlplane systemd[1]: Started kubelet: The Kubernetes Node Agent.
Jun 22 00:30:50 controlplane kubelet[1468]: Flag --network-plugin has been deprecated, will be removed along with dockershim.
Jun 22 00:30:50 controlplane kubelet[1468]: Flag --network-plugin has been deprecated, will be removed along with dockershim.
Jun 22 00:30:51 controlplane kubelet[1468]: I0622 00:30:51.059517    1468 server.go:446] "Kubelet version" kubeletVersion="v>
Jun 22 00:30:51 controlplane kubelet[1468]: I0622 00:30:51.060309    1468 server.go:874] "Client rotation is on, will bootst>
Jun 22 00:30:51 controlplane kubelet[1468]: E0622 00:30:51.088871    1468 certificate_manager.go:471] kubernetes.io/kube-api>
Jun 22 00:30:51 controlplane kubelet[1468]: I0622 00:30:51.089365    1468 dynamic_cafile_content.go:156] "Starting controlle>
Jun 22 00:30:51 controlplane kubelet[1468]: I0622 00:30:51.160243    1468 server.go:693] "--cgroups-per-qos enabled, but --c>
Jun 22 00:30:51 controlplane kubelet[1468]: I0622 00:30:51.160676    1468 container_manager_linux.go:281] "Container manager>
Jun 22 00:30:51 controlplane kubelet[1468]: I0622 00:30:51.161443    1468 container_manager_linux.go:286] "Creating Containe>
Jun 22 00:30:51 controlplane kubelet[1468]: I0622 00:30:51.162440    1468 topology_manager.go:133] "Creating topology manage>
Jun 22 00:30:51 controlplane kubelet[1468]: I0622 00:30:51.162460    1468 container_manager_linux.go:321] "Creating device p>
Jun 22 00:30:51 controlplane kubelet[1468]: I0622 00:30:51.165014    1468 state_mem.go:36] "Initialized new in-memory state >
Jun 22 00:30:51 controlplane kubelet[1468]: I0622 00:30:51.166366    1468 kubelet.go:313] "Using dockershim is deprecated, p>
Jun 22 00:30:51 controlplane kubelet[1468]: I0622 00:30:51.168693    1468 client.go:80] "Connecting to docker on the dockerE>
Jun 22 00:30:51 controlplane kubelet[1468]: I0622 00:30:51.168741    1468 client.go:99] "Start docker client with request ti>
Jun 22 00:30:51 controlplane kubelet[1468]: I0622 00:30:51.182659    1468 docker_service.go:571] "Hairpin mode is set but ku>
Jun 22 00:30:51 controlplane kubelet[1468]: I0622 00:30:51.182744    1468 docker_service.go:243] "Hairpin mode is set" hairp>
Jun 22 00:30:51 controlplane kubelet[1468]: I0622 00:30:51.183003    1468 cni.go:240] "Unable to update cni config" err="no >
Jun 22 00:30:51 controlplane kubelet[1468]: I0622 00:30:51.185283    1468 cni.go:240] "Unable to update cni config" err="no >
Jun 22 00:30:51 controlplane kubelet[1468]: I0622 00:30:51.185416    1468 docker_service.go:258] "Docker cri networking mana>
Jun 22 00:30:51 controlplane kubelet[1468]: I0622 00:30:51.185585    1468 cni.go:240] "Unable to update cni config" err="no >
Jun 22 00:30:51 controlplane kubelet[1468]: I0622 00:30:51.193297    1468 docker_service.go:264] "Docker Info" dockerInfo=&{>
Jun 22 00:30:51 controlplane kubelet[1468]: I0622 00:30:51.193330    1468 docker_service.go:279] "Setting cgroupDriver" cgro>
Jun 22 00:30:51 controlplane kubelet[1468]: I0622 00:30:51.216388    1468 kubelet.go:416] "Attempting to sync node with API >
Jun 22 00:30:51 controlplane kubelet[1468]: I0622 00:30:51.216421    1468 kubelet.go:278] "Adding static pod path" path="/et>
Jun 22 00:30:51 controlplane kubelet[1468]: I0622 00:30:51.216477    1468 kubelet.go:289] "Adding apiserver pod source"
Jun 22 00:30:51 controlplane kubelet[1468]: I0622 00:30:51.216508    1468 apiserver.go:42] "Waiting for node sync before wat>
Jun 22 00:30:51 controlplane kubelet[1468]: W0622 00:30:51.235161    1468 reflector.go:324] k8s.io/client-go/informers/facto>
Jun 22 00:30:51 controlplane kubelet[1468]: E0622 00:30:51.237684    1468 reflector.go:138] k8s.io/client-go/informers/facto>
Jun 22 00:30:51 controlplane kubelet[1468]: W0622 00:30:51.240097    1468 reflector.go:324] k8s.io/client-go/informers/facto>
Jun 22 00:30:51 controlplane kubelet[1468]: E0622 00:30:51.240182    1468 reflector.go:138] k8s.io/client-go/informers/facto>
Jun 22 00:30:51 controlplane kubelet[1468]: I0622 00:30:51.250132    1468 kuberuntime_manager.go:248] "Container runtime ini>
Jun 22 00:30:51 controlplane kubelet[1468]: W0622 00:30:51.250999    1468 probe.go:268] Flexvolume plugin directory at /usr/>
Jun 22 00:30:51 controlplane kubelet[1468]: I0622 00:30:51.252848    1468 server.go:1231] "Started kubelet"
```

Figure 9.19 – The kubelet service detailed logs

Then, it's up to you to find out what the main issue in the logs is. The following shows an example of the problem statement:

```
Jun 22 00:30:37 host01 systemd[1]: Started kubelet: The Kubernetes Node Agent.
Jun 22 00:30:39 controlplane kubelet[863]: E0622 00:30:39.592428     863 server.go:205] "Failed to load kubelet config file"
Jun 22 00:30:39 controlplane systemd[1]: kubelet.service: Main process exited, code=exited, status=1/FAILURE
Jun 22 00:30:39 controlplane systemd[1]: kubelet.service: Failed with result 'exit-code'.
```

Figure 9.20 – A sample kubelet agent error in the logs

Refer to *Chapter 6, Securing Kubernetes*, to learn about how to organize cluster access using `kubeconfig`. Once you have fixed the issue, you should restart the `kubelet` agent using the following command:

```
systemctl restart kubelet
```

Note that in the CKA exam, sometimes there isn't any real issue. After you have checked on the lost logs using the `journalctl -u kubelet.service` command, you could use some help from `systemctl restart kubelet` to reboot the `kubelet` agent to fix the issue.

Aside from issues with the cluster components, we often encounter application failures, the latter perhaps more often in the daily routine of working with Kubernetes clusters. So, let's now take a look at troubleshooting applications.

Troubleshooting applications

In this section, we'll focus on troubleshooting containerized applications deployed on the Kubernetes cluster. This commonly covers issues with containerized-application-related Kubernetes objects, including pods, containers, services, and StatefulSets. The troubleshooting skill that you will learn in this section will be helpful throughout your CKA exam.

Getting a high-level view

To troubleshoot the application failures, we have to start by getting a high-level view. The following command is the best way to get all the information at once:

```
kubectl get pods --all-namespaces
```

Alternatively, we can use the following:

```
kubectl get pods -A
```

The following output shows the pods up and running per namespace, within which you can easily find which pods have failed:

NAMESPACE	NAME	READY	STATUS	RESTARTS	AGE
app	old-busybox	0/1	ImagePullBackOff	0	2m54s
default	busybox	0/1	ImagePullBackOff	0	5m22s
default	nginx	1/1	Running	2 (3d7h ago)	10d
default	nginx-v2	1/1	Running	2 (3d7h ago)	10d
default	old-busybox	0/1	ImagePullBackOff	0	4m19s
default	test	1/1	Running	3 (3d7h ago)	10d
kube-system	coredns-6d4b75cb6d-4h89j	1/1	Running	2 (3d7h ago)	10d
kube-system	coredns-6d4b75cb6d-kj6cq	1/1	Running	2 (3d7h ago)	11d
kube-system	etcd-docker-desktop	1/1	Running	2 (3d7h ago)	11d
kube-system	kube-apiserver-docker-desktop	1/1	Running	2 (3d7h ago)	11d
kube-system	kube-controller-manager-docker-desktop	1/1	Running	2 (3d7h ago)	11d
kube-system	kube-flannel-ds-p7rhw	0/1	CrashLoopBackOff	327 (113s ago)	27h
kube-system	kube-proxy-9rfxs	1/1	Running	2 (3d7h ago)	11d
kube-system	kube-scheduler-docker-desktop	1/1	Running	2 (3d7h ago)	11d
kube-system	storage-provisioner	1/1	Running	4 (3d7h ago)	11d
kube-system	vpnkit-controller	1/1	Running	906 (2m56s ago)	11d

Figure 9.21 – Listing pods per namespace

To get the most out of the output information, note the NAMESPACE, READY, and STATUS columns – they will tell you in which namespace pods are up and running and how many copies. If you're certain about the failures that are happening on certain pods in a certain namespace, then you can move on to the next section to inspect the namespace events.

Inspecting namespace events

To inspect the namespace events, you can use the following command to find out what happened to the applications that were deployed in the `default` namespace:

```
kubectl get events
```

The output should look as follows:

```
LAST SEEN   TYPE      REASON         OBJECT                              MESSAGE
89s         Normal    Killing        pod/nginx-8f458dc5b-p74rr           Stopping container nginx
89s         Warning   FailedKillPod  pod/nginx-8f458dc5b-p74rr           error killing pod: failed to "KillContainer" for "nginx" wi
th KillContainerError: "rpc error: code = Unknown desc = Error response from daemon: No such container: b1f1e1318f8470f958506a2a4909b
b77e282e65f36a8bc0223ec9eacbb820db4"
89s         Normal    Killing        pod/nginx-8f458dc5b-v8j74           Stopping container nginx
89s         Warning   FailedKillPod  pod/nginx-8f458dc5b-v8j74           error killing pod: failed to "KillContainer" for "nginx" wi
th KillContainerError: "rpc error: code = Unknown desc = Error response from daemon: No such container: b94ddacdd8fa6b35e3ce66907ce32
e4924b297d0ffacf2d1d872d18e09f73c4c"
17s         Normal    Killing        pod/webfront-app-5c474b5bcc-nsvtt   Stopping container nginx
17s         Normal    Killing        pod/webfront-app-5c474b5bcc-zrwq2   Stopping container nginx
```

Figure 9.22 – The Kubernetes events

Within the preceding screenshot, we have some valuable columns:

- The `TYPE` column shows the event type – it could be `Normal` or `Warning`.
- The `REASON` column is tied to the behaviors of the events.
- The `OBJECT` column shows to which object this event is attached.
- The `MESSAGE` column shows what happened to a specific pod or container.

To know more about events, check out this blog to help you extract value from the Kubernetes event feed: `https://www.cncf.io/blog/2021/12/21/extracting-value-from-the-kubernetes-events-feed/`.

You can also sort the `events` list by most recent by using the following command:

```
kubectl get events --sort-by=.metadata.creationTimestamp
```

It will return the events sorted by their creation timestamp as follows:

```
LAST SEEN   TYPE      REASON    OBJECT                       MESSAGE
16m         Normal    Pulled    pod/vpnkit-controller        Container image "docker/desktop-vpnkit-controller:v2.0" already present on machine
24m         Normal    Created   pod/vpnkit-controller        Created container vpnkit-controller
24m         Normal    Started   pod/vpnkit-controller        Started container vpnkit-controller
16m         Warning   BackOff   pod/vpnkit-controller        Back-off restarting failed container
51s         Warning   BackOff   pod/kube-flannel-ds-p7rhw    Back-off restarting failed container
35m         Normal    Pulled    pod/kube-flannel-ds-p7rhw    Container image "rancher/mirrored-flannelcni-flannel:v0.18.1" already present on machine
```

Figure 9.23 – The Kubernetes events by timestamp

Similarly, if we wanted to check out the events in a namespace called `app`, we could use the following command:

```
kubectl get events -n app --sort-by=.metadata.creationTimestamp
```

The output should look as follows:

Figure 9.24 – The Kubernetes events per namespace by timestamp

The preceding output proves that we're able to print out the events per namespace and sort them by creation time stamp.

Up until this point, we're certain about which pod or container the issue occurred in. Now, let's take a closer look at the failing pods.

Troubleshooting failing pods

Once we narrow things down to the point where we know which pod is failing, we can use a command to get the pod status running in that namespace. The following is the command to get a failing pod called `old-busybox` in a namespace called `app`:

```
kubectl get pod old-busybox -n app
```

Your output will be similar to the following:

NAME	READY	STATUS	RESTARTS	AGE
old-busybox	0/1	ErrImagePull	0	106s

Figure 9.25 – Getting the failing pod in the namespace

We may notice that the STATUS shows there is an image error (ErrImagePull). Now, we can use the kubectl describe pod command to get more details:

```
kubectl describe pod old-busybox -n app
```

The preceding command prints an overview of the failing part, as shown in the following screenshot:

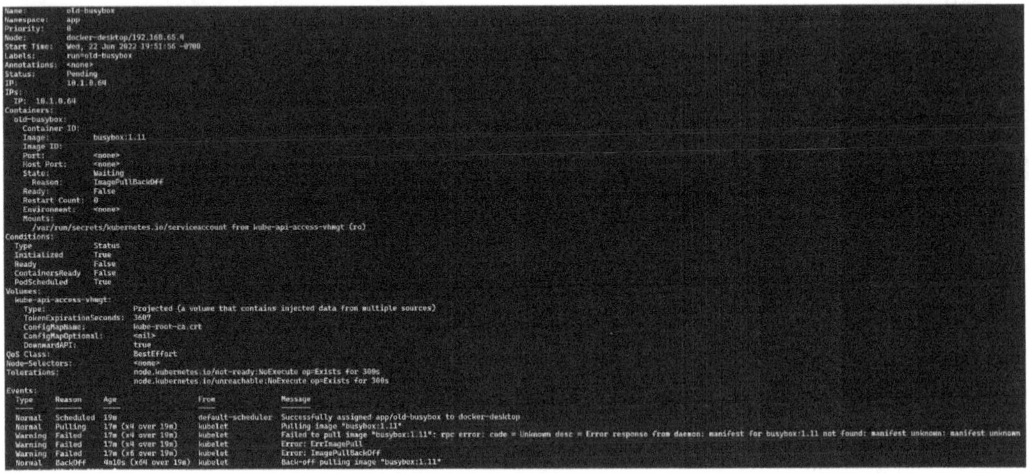

Figure 9.26 – Describing the failing pods in a namespace

You may notice there is a section called Events where the events related to this pod are displayed as follows:

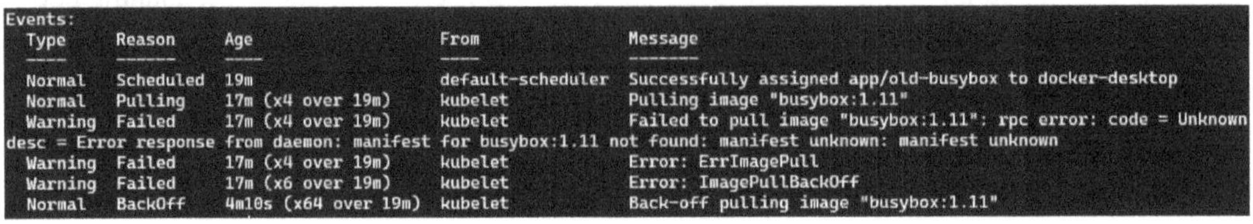

Figure 9.27 – The failing pod events

We can also use kubectl logs to get some information about the erroneous pod and the output will give you more detailed information. Let's use the same example to get the logs of a pod called old-busybox, as shown in the following command:

```
kubectl logs old-busybox -n app
```

The output is the following:

```
Error from server (BadRequest): container "old-busybox" in pod
"old-busybox" is waiting to start: trying and failing to pull
image
```

From the previous few outputs, we know the image was not correct. As this is a pod, we can use the following command to export the pod definition to a yaml file called my-old-pod.yaml:

```
kubectl get pod old-busybox -n app -o yaml > my-old-pod.yaml
```

We can also examine the content of this yaml file using the following command:

```
cat my-old-pod.yaml
```

The preceding command gives us the full configuration of the pod called old-busybox. However, we found the key part of this file is the section called image, as shown in the following:

```
apiVersion: v1
kind: Pod
metadata:
  creationTimestamp: "2022-06-23T03:33:40Z"
  labels:
    run: old-busybox
  name: old-busybox
  namespace: app
  resourceVersion: "812513"
  uid: c853b2ba-2f7e-4d26-806c-21d3ee7d6952
spec:
  containers:
  - image: busybox:1.11
    imagePullPolicy: IfNotPresent
    name: old-busybox
    resources: {}
    terminationMessagePath: /dev/termination-log
    terminationMessagePolicy: File
    volumeMounts:
    - mountPath: /var/run/secrets/kubernetes.io/serviceaccount
      name: kube-api-access-bkgmk
      readOnly: true
  dnsPolicy: ClusterFirst
  enableServiceLinks: true
  nodeName: docker-desktop
  preemptionPolicy: PreemptLowerPriority
  priority: 0
  restartPolicy: Always
  schedulerName: default-scheduler
  securityContext: {}
  serviceAccount: default
  serviceAccountName: default
  terminationGracePeriodSeconds: 30
```

Figure 9.28 – The failing pod specification

We can edit this exported file locally using the following command:

```
vim my-old-pod.yaml
```

You'll see that you can edit the YAML file when you're in EDIT mode as follows:

```
apiVersion: v1
kind: Pod
metadata:
  creationTimestamp: "2022-06-23T03:33:40Z"
  labels:
    run: old-busybox
  name: old-busybox
  namespace: app
  resourceVersion: "812513"
  uid: c853b2ba-2f7e-4d26-806c-21d3ee7d6952
spec:
  containers:
  - image: busybox:latest
    imagePullPolicy: IfNotPresent
    name: old-busybox
    resources: {}
    terminationMessagePath: /dev/termination-log
    terminationMessagePolicy: File
```

Figure 9.29 – Editing the pod-exported YAML specification

After you're done with the editing, you need to delete the old pod using the kubectl delete command, as follows:

```
kubectl delete pod old-busybox -n app
```

Then, deploy my-old-pod using the kubectl apply -f command, and then you'll see the pod is up and running again:

```
NAME          READY   STATUS    RESTARTS      AGE
old-busybox   1/1     Running   3(36s ago)    51s
```

> **Important note**
>
> For a failing pod that was initiated by deployment, you can use kubectl edit deploy < your deployment > to live-edit the pod and fix the error. It helps to quickly fix a range of errors. To learn more about how the deployment live-edit works, refer to *Chapter 4, Application Scheduling and Lifecycle Management.*

The failing pods include the following cases:

Failing type	How to debug?
Pending	Use the `kubectl describe` command – sometimes, it is a scheduling issue because of no available nodes or exceeding the resource. Make sure you check the node status and use the `top` command to check out the resource allocation.
CrashLoopBackOff	Use the `kubectl describe` and `kubectl log` commands – sometimes, it was caused by cluster components, so make sure you narrow the error down by using the outside-in approach.
Completed	Use the `kubectl describe` command to find out why it happened and then fix it.
Error	Use the `kubectl describe` command to find out why it happened and then fix it.
ImagePullBackOff	`kubectl` describes and mostly needs to export the YAML file, then update the image. Also possible to use the `set image` command.

Table 9.2 - Failing pods and how to fix them

Knowing about pod troubleshooting comes in handy and applies to most cases, in particular in the microservices architecture where there is mainly one container per pod. When it comes to multiple containers in a pod or a pod containing init containers, we'll need to execute a command on the pod to troubleshoot – let's take a look at those cases now.

Troubleshooting init containers

In *Chapter 4, Application Scheduling and Lifecycle Management*, of this book, we learned about init containers, as we deployed init containers in the following example:

```
apiVersion: v1
kind: Pod
metadata:
  name: packt-pod
  labels:
    app: packtapp
```

```
spec:
  containers:
  - name: packtapp-container
    image: busybox:latest
    command: ['sh', '-c', 'echo The packtapp is running! &&
sleep 3600']
  initContainers:
  - name: init-packtsvc
    image: busybox:latest
    command: ['sh', '-c', 'until nslookup init-packtsvc; do
echo waiting for init-packtsvc;  sleep 2;  done;']
```

We can use the following command to check the status for the initContainer of this pod:

```
kubectl get pod packt-pod --template '{{. status.
initContainerStatuses}}'
```

In my case, the printed output looks as follows:

```
[map[containerID:docker://016f1176608e521b3eecde33c35dce3596
a46a483f38a69ba94ed48b8dd91f13 image:busybox:latest imageID:
docker-pullable://busybox@sha256:3614ca5eacf0a3a1bcc361c939202
a974b4902b9334ff36eb29ffe9011aaad83 lastState:map[] name:
init-packtsvc ready:false restartCount:0  state:map[running:map
[startedAt:2022-06-23T04:57:47Z]]]]
```

The preceding output shows that the initContainer is not ready.

We can use the following command to check the logs for the initContainer of the pod to understand why and fix the issue:

```
kubectl logs packt-pod -c init-packtsvc
```

Similarly, initContainer also has its status – the following are the common ones:

Failing type	What does that mean?
`Init: X/Y`	The pod has `Y` init containers in total and `X` of them are completed
`Init: Error`	`initContainer` failed to execute correctly
`Init:CrashLoopBackOff`	`initContainer` is failing repeatedly
`Pending`	The pod is pending, so it has not started the `initContainer` execution yet
`PodInitializing`	The `initContainer` is executed and now the pod is initiating
`Running`	The `initContainer` is executed and now the pod is up and running

Familiarity with these statuses will help you define when and how to take further steps to debug containers.

Summary

This chapter covered cluster troubleshooting and application troubleshooting from the cluster, the node, and then down to the pod level – this is an end-to-end, outside-in approach. As a Kubernetes administrator, acquiring good troubleshooting skills will help you to provide better value to your organization greatly.

In the next chapter, we'll focus on Kubernetes security, networking troubleshooting use cases, and some more end-to-end troubleshooting scenarios. Stay tuned!

FAQs

- *Where can I find a comprehensive guide to troubleshooting the clusters?*

  You can find the updated information from the official Kubernetes documentation:

  `https://kubernetes.io/docs/tasks/debug/debug-cluster/`

- *Where can I find a comprehensive guide to troubleshooting the applications?*

  You can find the updated information from the official Kubernetes documentation:

  `https://kubernetes.io/docs/tasks/debug/debug-application/`

10
Troubleshooting Security and Networking

So far in this book, we have talked about Kubernetes architecture, the application life cycle, security, and networking. I hope that since this is the last chapter, we can follow on from *Chapter 9, Troubleshooting Cluster Components and Applications*, to talk about security and networking troubleshooting. This chapter provides the general troubleshooting approaches for troubleshooting errors caused by RBAC restrictions or networking settings. We have touched upon how to enable Kubernetes RBAC in *Chapter 6, Securing Kubernetes*, and upon working with Kubernetes DNS in *Chapter 7, Demystifying Kubernetes Networking*. Be sure to go back to these chapters and review the important concepts before diving into this chapter. We're going to cover the following main topics in this chapter:

- Troubleshooting RBAC failures
- Troubleshooting networking

Technical requirements

To get started, we need to make sure your local machine meets the following technical requirements.

In case you're on Linux, we're demonstrating examples with a `minikube` cluster – check out *Chapter 2, Installing and Configuring Kubernetes Clusters*. Make sure that your test environment meets the following requirements:

- A compatible Linux host. We recommend a Debian-based Linux distribution such as Ubuntu 18.04 or later.
- Make sure that your host machine has at least 2 GB of RAM, 2 CPU cores, and about 20 GB of free disk space.

In case you're on Windows 10 or Windows 11, make note of the following:

- We recommend updating Docker Desktop to the latest version and creating a local `docker-desktop` Kubernetes cluster. Refer to this article to understand how to set up a local Kubernetes cluster with Docker Desktop: `https://docs.docker.com/desktop/kubernetes/`.
- We also recommend using **Windows Subsystem for Linux 2** (**WSL 2**) to test the environment – refer to this article to see how to install WSL 2 (`https://docs.microsoft.com/en-us/windows/wsl/install`) and the following article to see how to set up the Docker Desktop WSL 2 backend (`https://docs.docker.com/desktop/windows/wsl/`).

Troubleshooting RBAC failures

Troubleshooting any issues related to Kubernetes security seems a bit contradictory. As a matter of fact, most of the security layers of Kubernetes involve working with tooling that helps secure the 4C layers of Kubernetes, which involves security scanning, managing, and protection. To learn more about the 4C layers, please refer to *Chapter 6, Securing Kubernetes*. When it comes to troubleshooting security, the CKA exam is most often about the Kubernetes RBAC issue. Therefore, we'll focus on showing an example of troubleshooting RBAC failures in Kubernetes in this section.

Initiating a minikube cluster

This part is not covered by the CKA exam, but you may encounter this if you're trying to deploy the `minikube` cluster by yourself following the instructions in *Chapter 2, Installing and Configuring the Kubernetes Cluster*. You will need to apply what we discussed in that chapter of the book whenever you're trying to install a new `minikube` cluster in a virgin Linux VM.

After you have installed the `minikube` tools, you can start to spin up your local cluster using the following command:

```
minikube start
```

You may see the following error in your output:

```
👎  Unable to pick a default driver. Here is what was considered, in preference order:
    ▪ docker: Not healthy: "docker version --format {{.Server.Os}}-{{.Server.Version}}" exit status 1:
    ▪ docker: Suggestion:  <https://minikube.sigs.k8s.io/docs/drivers/docker/>
💡  Alternatively you could install one of these drivers:
    ▪ kvm2: Not installed: exec: "virsh": executable file not found in $PATH
    ▪ podman: Not installed: exec: "podman": executable file not found in $PATH
    ▪ vmware: Not installed: exec: "docker-machine-driver-vmware": executable file not found in $PATH
    ▪ virtualbox: Not installed: unable to find VBoxManage in $PATH
    ▪ qemu2: Not installed: exec: "qemu-system-x86_64": executable file not found in $PATH

❌  Exiting due to DRV_NOT_HEALTHY: Found driver(s) but none were healthy. See above for suggestions how to fix installed drivers.
```

Figure 10.1 – The drivers are not healthy

Your first instinct is to choose the correct drive and use the `sudo` command, as in the following:

```
sudo minikube start --driver=docker
```

As a result, you may see the following output:

```
cloudmelon@cloudmelonsrv:~$ sudo minikube start --driver=docker
⊜ minikube v1.24.0 on Ubuntu 18.04
✳ Using the docker driver based on user configuration
⊛ The "docker" driver should not be used with root privileges.
❗ If you are running minikube within a VM, consider using --driver=none:
▮    https://minikube.sigs.k8s.io/docs/reference/drivers/none/

✘ Exiting due to DRV_AS_ROOT: The "docker" driver should not be used with root privileges.
```

Figure 10.2 – The service account per namespace

The preceding output was because of the Docker root privileges issue. The best practice is to manage Docker as a non-root user to avoid this issue. In order to achieve this, we need to add a user to a group called `docker`:

1. Create the `docker` group:

```
sudo groupadd docker
```

2. Add your user to the group called `docker`:

```
sudo usermod -aG docker $USER
```

3. From here you need to log in again or restart the Docker server so that your group membership is re-evaluated. However, we should activate the changes to the group by using the following command when we're on the Linux OS:

```
newgrp docker
```

4. The next time, when you log in, use the following command if you want Docker to start on boot:

```
sudo systemctl enable docker.service
sudo systemctl enable containerd.service
```

5. After the preceding steps, you should be able to start `minikube` with the Docker driver by using the following command:

```
minikube start --driver=docker
```

The preceding `minikube start` command has created a `minikube` cluster successfully if you are able to see an output similar to the following:

```
minikube v1.25.2 on Ubuntu 22.04
minikube 1.26.0 is available! Download it: https://github.com/kubernetes/minikube/releases/tag/v1.26.0
To disable this notice, run: 'minikube config set WantUpdateNotification false'

Using the docker driver based on existing profile
Starting control plane node minikube in cluster minikube
Pulling base image ...
Restarting existing docker container for "minikube" ...
This container is having trouble accessing https://k8s.gcr.io
To pull new external images, you may need to configure a proxy: https://minikube.sigs.k8s.io/docs/reference/networking/proxy/
Preparing Kubernetes v1.23.3 on Docker 20.10.12 ...
  • kubelet.housekeeping-interval=5m
Verifying Kubernetes components...
  • Using image gcr.io/k8s-minikube/storage-provisioner:v5
Enabled addons: storage-provisioner, default-storageclass
Done! kubectl is now configured to use "minikube" cluster and "default" namespace by default
```

Figure 10.3 – Starting the minikube cluster successfully

Although this section is not covered in the CKA exam, it's highly recommended to get familiar with it in case you're stuck when creating a `minikube` cluster. Once you get your `minikube` cluster up and running, we can get into managing a `minikube` cluster and troubleshooting RBAC as needed.

Managing a minikube cluster

When it comes to managing a `minikube` cluster, we learned in *Chapter 6*, *Securing Kubernetes*, that we need to set `apiserver --authorization-mode` to RBAC in order to enable Kubernetes RBAC, as shown in the following example:

```
kube-apiserver --authorization-mode=RBAC
```

Make sure that our current context uses our default `minikube` and then use the following commands to create a new deployment in a specific namespace:

```
kubectl create ns app
kubectl create deployment rbac-nginx --image=nginx -n app
```

The preceding two commands create a namespace called `app`, and a new deployment called `rbac-nginx` within the `app` namespace.

Let's define a new role called `rbac-user` in a namespace called `app` by using the following command:

```
kubectl create role rbac-user --verb=get --verb=list
--resource=pods --namespace=app
```

We then need to create rolebinding to bind this role to the subjects, as is shown in the following command:

```
kubectl create rolebinding rbac-pods-binding --role=rbac-user
--user=rbac-dev --namespace=app
```

As `rbac-user` only has to list and get permissions for pods, let's try to use this profile for user impersonation to delete the deployment:

```
kubectl auth can-i delete deployment --as=rbac-user
```

The output should look as follows:

```
No
```

You can learn more about user impersonation from the official documentation here: `https://kubernetes.io/docs/reference/access-authn-authz/authentication/#user-impersonation`

To resolve the issue, we could update the role for `rbac-user` in the YAML definition, as in the following:

```
apiVersion: rbac.authorization.k8s.io/v1
kind: Role
metadata:
  namespace: app
  name: rbac-user
rules:
- apiGroups: ["extensions", "apps"]
  resources: ["deployments"]
  verbs: ["get", "list", "watch", "create", "update", "patch",
"delete"]
```

We could use the `kubectl auth reconcile` command to create or update a YAML manifest file containing RBAC objects. Check the official documentation for more information (`https://kubernetes.io/docs/reference/access-authn-authz/rbac/#kubectl-auth-reconcile`):

```
kubectl auth reconcile -f my-rbac-rules.yaml
```

The RBAC issue applies in the use case where different dev teams are sharing the cluster resources – as a Kubernetes administrator, you're likely to access the cluster with full permission. Understanding this part will help you better govern the permissions among the dev team members for a better standard of security and compliance.

Troubleshooting networking

In *Chapter 7, Demystifying Kubernetes Networking*, we learned that the Kubernetes DNS server creates DNS records (A/AAAA, SRV, and PTR records) for services and pods in Kubernetes. Those efforts allow you to contact Services with consistent DNS names in place of the IP addresses. The Kubernetes DNS server does this by scheduling a few copies of DNS pods and services on the Kubernetes cluster.

In the following section, let's talk about how to troubleshoot the Kubernetes DNS service.

Troubleshooting a Kubernetes DNS server

To troubleshoot the networking of Kubernetes, we start by checking the status of the DNS server. Using `minikube` as a local cluster this time, we use the following command to check whether the DNS server is up and running on your cluster:

```
kubectl get pods -n kube-system | grep dns
```

The output should be similar to the following:

```
coredns-64897985d-brqfl 1/1 Running 1 (2d ago) 2d
```

From the preceding output, we can see that the CoreDNS is up and running in our current `minikube` cluster. We can also do this by using the `kubectl get deploy core-dns -n kube-system` command.

To get further details, we check out the CoreDNS deployment settings by using the `kubectl describe` command, as in the following:

```
kubectl describe deploy coredns -n kube-system
```

The output is as follows:

```
Name:                   coredns
Namespace:              kube-system
CreationTimestamp:      Mon, 14 Mar 2022 00:25:33 +0000
Labels:                 k8s-app=kube-dns
Annotations:            deployment.kubernetes.io/revision: 1
Selector:               k8s-app=kube-dns
Replicas:               1 desired | 1 updated | 1 total | 1 available | 0 unavailable
StrategyType:           RollingUpdate
MinReadySeconds:        0
RollingUpdateStrategy:  1 max unavailable, 25% max surge
Pod Template:
  Labels:               k8s-app=kube-dns
  Service Account:      coredns
  Containers:
   coredns:
    Image:        k8s.gcr.io/coredns/coredns:v1.8.6
    Ports:        53/UDP, 53/TCP, 9153/TCP
    Host Ports:   0/UDP, 0/TCP, 0/TCP
    Args:
      -conf
      /etc/coredns/Corefile
    Limits:
      memory:  170Mi
    Requests:
      cpu:     100m
      memory:  70Mi
    Liveness:     http-get http://:8080/health delay=60s timeout=5s period=10s #success=1 #failure=5
    Readiness:    http-get http://:8181/ready delay=0s timeout=1s period=10s #success=1 #failure=3
    Environment:  <none>
    Mounts:
      /etc/coredns from config-volume (ro)
  Volumes:
   config-volume:
    Type:               ConfigMap (a volume populated by a ConfigMap)
    Name:               coredns
    Optional:           false
  Priority Class Name:  system-cluster-critical
Conditions:
  Type           Status  Reason
  ----           ------  ------
  Available      True    MinimumReplicasAvailable
  Progressing    True    NewReplicaSetAvailable
OldReplicaSets:  <none>
NewReplicaSet:   coredns-64897985d (1/1 replicas created)
Events:          <none>
```

Figure 10.4 – The minikube CoreDNS configurations

As we said, the Kubernetes DNS service creates DNS records for services, so you contact services with a consistent DNS fully qualified hostnames instead of IP addresses. As it is located in the kube-system namespace, we can check it out by using the following command for our minikube cluster:

```
kubectl get svc -n kube-system
```

The output is as follows, which gives us the cluster IP of kube-dns:

```
NAME          TYPE      CLUSTER-IP    EXTERNAL-IP    PORT(S)
AGE
kube-dns      ClusterIP 10.96.0.10    <none>         53/UDP,53/TCP,9153/
TCP 2d
```

To troubleshoot issues with the DNS server, we can use the kubectl logs command:

```
kubectl logs coredns-64897985d-brqfl -n kube-system
```

The preceding kubectl logs command shows the logs for a coredns pod named coredns-64897985d-brqfl and the output is similar to the following:

```
[INFO] plugin/ready: Still waiting on: "kubernetes"
[INFO] plugin/ready: Still waiting on: "kubernetes"
[INFO] plugin/ready: Still waiting on: "kubernetes"
[WARNING] plugin/kubernetes: starting server with unsynced Kubernetes API
.:53
[INFO] plugin/reload: Running configuration MD5 = cec3c60eb1cc4909fd4579a8d79ea031
CoreDNS-1.8.6
linux/arm64, go1.17.1, 13a9191
[INFO] plugin/ready: Still waiting on: "kubernetes"
[INFO] plugin/ready: Still waiting on: "kubernetes"
```

Figure 10.5 – The minikube CoreDNS logs

The output shows whether the DNS server is up or not, and will log abnormal events if any exist. Once we know that the DNS server is up, we can take a look at how to troubleshoot the services deployed in the Kubernetes cluster in the following section.

Troubleshooting a service in Kubernetes

To troubleshoot a service, let's first deploy a new deployment called svc-nginx:

```
kubectl create deployment svc-nginx --image=nginx -n app
```

The following output shows that it has been created successfully:

```
deployment.apps/svc-nginx created
```

Let's now take a look at exposing a service for the svc-nginx deployment. We're using the following command to expose the NodePort service of the nginx pod on port 80:

```
kubectl expose deploy svc-nginx --type=NodePort --name=nginx-
svc --port 80 -n app
```

The following output shows that it has been exposed successfully:

```
service/nginx-svc exposed
```

As we learned from *Chapter 7, Demystifying Kubernetes Networking*, we know that we can expect the nginx-svc service to follow the general service DNS name pattern, which would be as follows:

```
nginx-svc.app.svc.cluster.local
```

Now, let's take a look at the services currently in the app namespace of our Kubernetes cluster by using the following command:

```
kubectl get svc -n app
```

We can see an output similar to the following:

```
cloudmelon@cloudmelonplayground:~$ kubectl get svc -n app
NAME        TYPE       CLUSTER-IP      EXTERNAL-IP   PORT(S)        AGE
nginx-svc   NodePort   10.101.34.154   <none>        80:32242/TCP   83s
```

Figure 10.6 – A nginx-svc service in the Kubernetes app namespace

From the preceding output, we can get a closer look at nginx-svc by using the following command:

```
kubectl get svc nginx-svc -n app -o wide
```

The output of the preceding command is the following:

```
cloudmelon@cloudmelonplayground:~$ kubectl get svc nginx-svc -n app -o wide
NAME        TYPE       CLUSTER-IP      EXTERNAL-IP   PORT(S)        AGE     SELECTOR
nginx-svc   NodePort   10.101.34.154   <none>        80:32242/TCP   2m42s   app=svc-nginx
```

Figure 10.7 – A closer look at the nginx-svc service

The preceding command shows that the IP address of the nginx-svc service is 10.101.34.154, so let's use the nslookup command to check out its DNS name:

```
kubectl run -it sandbox --image=busybox:latest --rm
--restart=Never -- nslookup 10.101.34.154
```

> **Important Note**
> The preceding command creates a `busybox` pod in the default namespace. As by default, pods in the Kubernetes cluster can talk to each other, we could use a `sandbox` pod to test the connectivity to a different namespace.

The preceding command will give you the following output:

```
Server:          10.96.0.10
Address:         10.96.0.10:53

154.34.101.10.in-addr.arpa         name = nginx-svc.app.svc.cluster.local

pod "sandbox" deleted
```

Figure 10.8 – Returning back the DNS name for nginx-svc

If you want to test the connectivity by using a pod in the same namespace as `nginx-svc`, use the following command:

```
kubectl run -it sandbox -n app --image=busybox:latest --rm
--restart=Never -- nslookup 10.101.34.154
```

Based on the preceding output, we can see the DNS name for `nginx-svc` is `nginx-svc.app.svc.cluster.local`. Now, let's get the DNS record of the `nginx-svc` service from the `app` namespace using the following command:

```
kubectl run -it sandbox --image=busybox:latest --rm
--restart=Never -- nslookup nginx-svc.app.svc.cluster.local
```

You'll see that the output is similar to the following:

```
Server:     10.96.0.10
Address 1:  10.96.0.10 kube-dns.kube-system.svc.cluster.local

Name:       nginx-svc.app.svc.cluster.local
Address 1:  10.101.34.154 nginx-svc.app.svc.cluster.local
pod "sandbox" deleted
```

Now, let's test out the connectivity of the `nginx-svc` service. We can use the `nginx-beta` deployment to see what's coming back using `curl`. The complete command is as follows:

```
kubectl run -it nginx-beta -n app --image=nginx --rm
--restart=Never -- curl -Is http://nginx-svc.app.svc.cluster.
local
```

The output is as follows:

```
If you don't see a command prompt, try pressing enter.

HTTP/1.1 200 OK
Server: nginx/1.23.0
Date: Sun, 26 Jun 2022 22:40:05 GMT
Content-Type: text/html
Content-Length: 615
Last-Modified: Tue, 21 Jun 2022 14:25:37 GMT
Connection: keep-alive
ETag: "62b1d4e1-267"
Accept-Ranges: bytes

pod "nginx-beta" deleted
```

Figure 10.9 – Returning the nginx main page

The preceding screenshot with 200 responses proves that the connectivity between the nginx-beta pod and the nginx-svc Service is OK, and that we managed to use curl on the main page of nginx with the DNS name of the nginx service.

The approach that we discussed in this section works well when we want to quickly test the connectivity within the same namespace or to a different namespace. The latter would also work in a scenario where the network policy is deployed to restrict the connectivity between pods in different namespaces. Now, in the following section, let's take a look at how to get a shell to debug the Kubernetes networking in case we need a longer session.

Get a shell for troubleshooting

Given the same scenario with the svc-nginx deployment in the app namespace, now let's use the interactive shell to troubleshoot the networking.

After we find the IP address of nginx-svc, 10.101.34.154, let's use the nslookup command to check out its DNS name – use the following command:

```
kubectl run -it sandbox --image=busybox:latest --rm
--restart=Never --
```

We're now getting into the interactive shell:

```
If you don't see a command prompt, try pressing enter.
/ # whoami
root
```

In this interactive shell, we log in as root, and we can use `nslookup` or another valid command to troubleshoot the networking:

```
nslookup 10.101.34.154
```

The output is as follows:

```
/ # whoami
root
/ # nslookup 10.101.34.154
Server:         10.96.0.10
Address:        10.96.0.10:53

154.34.101.10.in-addr.arpa      name = nginx-svc.app.svc.cluster.local
```

Figure 10.10 – An interactive shell in BusyBox

There's a handful of commands available in BusyBox, though `curl` isn't one of them. So, let's now get an `nginx` image with `curl` available. To know what the shell commands available in BusyBox are, refer to the following page: `https://hub.docker.com/_/busybox`.

We can use the following command to get to the interactive shell of the `nginx` pod and find the `nginx` pod:

```
kubectl get pods -n app | grep svc-nginx
```

Then, it will come back with the full name of the pod that the `svc-nginx` deployment created:

```
svc-nginx-77cbfd944c-9wp6s     1/1        Running      0
4h14m
```

Let's use the `kubectl exec` command to get the interactive shell:

```
kubectl exec -i -t svc-nginx-77cbfd944c-9wp6s --container nginx
-n app -- /bin/bash
```

The preceding command will get you the interactive shell access, and then we can use the same `curl` command to test the connectivity:

```
root@svc-nginx-77cbfd944c-9wp6s:/#
curl -Is http://nginx-svc.app.svc.cluster.local
```

This technique comes in extremely handy in a case where a pod has one or more containers. Refer to this article to get more tips: `https://kubernetes.io/docs/tasks/debug/debug-application/get-shell-running-container/`.

In this section, we have covered troubleshooting networking – the commands presented in this section are references that you can leverage in your real-life debugging session. Go back and practice a few times, make sure you get a proper understanding, and it will pay off.

Summary

This chapter has covered the approaches and use cases for Kubernetes RBAC and networking troubleshooting. Together with *Chapter 8, Monitoring and Logging Kubernetes Clusters and Applications*, and *Chapter 9, Troubleshooting Cluster Components and Applications*, that covers 30% of the CKA content.

To get the most out of this chapter, go back and refer to *Chapter 6, Securing Kubernetes*, especially the section on how to enable Kubernetes RBAC, and to *Chapter 7, Demystifying Kubernetes*, to refresh how to work with Kubernetes DNS. Knowing how to work with Kubernetes DNS will help you lay the foundations for understanding other important concepts.

Make sure that you check out the *FAQs* section in all the chapters for further references, as well as reading all the recommended documentation and articles. A good understanding of these materials will help you become more confident in your daily job as a Kubernetes administrator.

Let's stay tuned!

FAQs

- *Where can I find a comprehensive guide to troubleshooting the Kubernetes services?*

  You can find the updated documentation within the official Kubernetes documentation:

  ```
  https://kubernetes.io/docs/tasks/debug/debug-application/debug-
  service/
  ```

  Also highly recommended is focusing on this chapter together with *Chapter 9, Troubleshooting Cluster Components and Applications*, as a complementary resource. This will help you gather a full view of the Kubernetes troubleshooting story.

- *Where can I find a comprehensive guide to Kubernetes networking?*

  Chapter 7 of this book, *Demystifying Kubernetes Networking*, touches upon most of the Kubernetes networking concepts, as well as troubleshooting examples – together with this chapter, this will help you work confidently on questions that could appear in the actual CKA exam. You can also bookmark the following article from the official Kubernetes documentation:

  ```
  https://kubernetes.io/docs/concepts/cluster-administration/
  networking/
  ```

Appendix - Mock CKA scenario-based practice test resolutions

Chapter 2 – Installing and Configuring Kubernetes Clusters

You have two virtual machines: master-0 and worker-0. Please complete the following mock scenarios.

Scenario 1

Install the latest version of kubeadm, then create a basic kubeadm cluster on the master-0 node, and get the node information.

1. Update the apt package index, add a Google Cloud public signing key, and set up the Kubernetes apt repository by running the following instructions:

```
sudo apt-get update
sudo apt-get install -y apt-transport-https
ca-certificates curl
sudo curl -fsSLo /usr/share/keyrings/kubernetes-archive-
keyring.gpg https://packages.cloud.google.com/apt/doc/
apt-key.gpg
echo "deb [signed-by=/usr/share/keyrings/kubernetes-
archive-keyring.gpg] https://apt.kubernetes.io/
kubernetes-xenial main" | sudo tee /etc/apt/sources.
list.d/kubernetes.list
```

2. Start by updating the apt package index, then install kubelet and kubeadm:

```
sudo apt-get update
sudo apt-get install -y kubelet kubeadm
```

3. At this point, if you haven't installed kubectl yet, you can also install kubelet, kubeadm, and kubectl in one go:

```
sudo apt-get update
sudo apt-get install -y kubelet kubeadm kubectl
```

4. Use the following command to pin the version of the utilities you're installing:

```
sudo apt-mark hold kubelet kubeadm kubectl
```

5. You can use the kubeadm init command to initialize the control-plane like a regular user, and gain sudo privileges from your master node machine by using the following command:

```
sudo kubeadm init --pod-network-cidr=192.168.0.0/16
```

6. After your Kubernetes control-plane is initialized successfully, you can execute the following commands to configure kubectl:

```
mkdir -p $HOME/.kube
sudo cp -i /etc/kubernetes/admin.conf $HOME/.kube/config
sudo chown $(id -u):$(id -g) $HOME/.kube/config
```

Scenario 2

SSH to worker-0 and join it to the master-0 node.

You can use the following command to join the worker nodes to the Kubernetes cluster. This command can be used repeatedly each time you have new worker nodes to join with the token that you acquired from the output of the kubeadm control-plane:

```
sudo kubeadm join --token <token>  <control-plane-
host>:<control-plane-port> --discovery-token-ca-cert-hash
sha256:<hash>
```

Scenario 3 (optional)

Set up a local minikube cluster, and schedule your first workload called `hello Packt`.

> **Note**
>
> Check out the *Installing and configuring Kubernetes cluster* section in *Chapter 2*, to set up a single node minikube cluster.

Let's quickly run an app on the cluster called `helloPackt` using `busybox`:

```
kubectl run helloPackt --image=busybox
```

Chapter 3 – Maintaining Kubernetes Clusters

You have two virtual machines: `master-0` and `worker-0`. Please complete the following mock scenarios.

Scenario 1

SSH to the `master-0` node, check the current `kubeadm` version, and upgrade to the latest `kubeadm` version. Check the current `kubectl` version, and upgrade to the latest `kubectl` version.

Start by checking the current version with the following commands once we're in the master node:

```
kubeadm version
kubectl version
```

Check out the latest available versions:

```
apt update
apt-cache madison kubeadm
```

Upgrade the `kubeadm` using the following command:

```
apt-mark unhold kubeadm && \
apt-get update && apt-get install -y kubeadm=1.xx.x-00 && \
apt-mark hold kubeadm
```

Check if your cluster can be upgraded, and the available versions that your cluster can be upgraded to by using the following command:

```
kubeadm upgrade plan
```

Use the following command to upgrade the kubeadm:

```
kubeadm upgrade apply v1.xx.y
```

Scenario 2

SSH to worker-0 node, check the current kubeadm version, and upgrade to the latest kubeadm version. Check the current kubelet version, and upgrade to the latest kubelet version.

Start by checking the current version with the following commands once we're in the master node:

```
kubeadm version
kubectl version
```

Check what the latest versions available are:

```
apt update
apt-cache madison kubeadm
```

Upgrade the kubelet (which also upgrades the local kubelet configuration) with the following command:

```
sudo kubeadm upgrade node
```

Cordon the node so that we drain the workloads of preparing the node for maintenance using the following command:

```
kubectl drain worker-0 --ignore-daemonsets
```

Upgrade the kubeadm by using the following command:

```
apt-mark unhold kubeadm && \
apt-get update && apt-get install -y kubeadm=1.xx.x-00 && \
apt-mark hold kubeadm
```

Check if your cluster can be upgraded and the available versions that your cluster can be upgraded to by using the following command:

```
kubeadm upgrade plan
```

Use the following command to upgrade the kubeadm:

```
kubeadm upgrade apply v1.xx.y
```

Restart the `kubelet` for the changes to take effect:

```
sudo systemctl daemon-reload
sudo systemctl restart kubelet
```

Finally, we can uncordon the worker node and it will return the node that is now shown as uncordoned:

```
kubectl uncordon worker-0
```

Scenario 3

SSH to the `master-0` node, and backup the `etcd` store.

Use the following command to check the endpoint status:

```
sudo ETCDCTL_API=3 etcdctl endpoint status --endpoints=ht
tps://172.16.16.129:2379 --cacert=/etc/kubernetes/pki/etcd/
ca.crt --cert=/etc/kubernetes/pki/etcd/server.crt --key=/etc/
kubernetes/pki/etcd/server.key --write-out=table
```

Use the following command to backup `etcd`:

```
sudo ETCDCTL_API=3 etcdctl snapshot save snapshotdb
--endpoints=https://172.16.16.129:2379
--cacert=/etc/kubernetes/pki/etcd/ca.crt --cert=/etc/
kubernetes/pki/etcd/server.crt --key=/etc/kubernetes/pki/etcd/
server.key
```

Scenario 4

SSH to the `master-0` node, and restore the `etcd` store to the previous backup.

Restore the `etcd` from a previous backup operation using the following command:

```
sudo ETCDCTL_API=3 etcdctl --endpoints 172.16.16.129:2379
snapshot restore snapshotdb
```

Chapter 4 – Application scheduling and lifecycle management

You have two virtual machines: master-0 and worker-0, please complete the following mock scenarios.

Scenario 1

SSH to the worker-0 node, and provision a new pod called ngnix with a single container nginx.

Use the following command:

```
kubectl run nginx --image=nginx:alpine
```

Scenario 2

SSH to worker-0, and then scale the nginx to 5 copies.

Use the following command:

```
kubectl scale deployment nginx --replicas=5
```

Scenario 3

SSH to worker-0, set a configMap with a username and password, then attach a new pod with a busybox.

Create a yaml definition called packt-cm.yaml to define ConfigMap as the following:

```
apiVersion: v1
kind: ConfigMap
metadata:
  name: packt-configmap
data:
  myKey: packtUsername
  myFav: packtPassword
```

Use the following command to deploy the yaml manifest:

```
kubectl apply -f packt-cm.yaml
```

Verify the `configMap` by using the following command:

```
kubectl get configmap
```

Once you have `configMap` ready, create a yaml definition file to config the pod to consume the `configMap` as the following:

```
apiVersion: v1
kind: Pod
metadata:
  name: packt-configmap
spec:
  containers:
  - name: packt-container
    image: busybox
    command: ['sh', '-c', "echo $(MY_VAR) && sleep 3600"]
    env:
    - name: MY_VAR

      valueFrom:
        configMapKeyRef:
          name: packt-configmap
          key: myKey
```

Use the following command to verify the `configMap` value:

```
kubectl logs packt-configmap
```

Scenario 4

SSH to `worker-0`, and create a nginx pod with an `initContainer` called `busybox`.

Create a yaml definition called `packt-pod.yaml` shown as follows:

```
apiVersion: v1
kind: Pod
metadata:
  name: packtpod
  labels:
    app: packtapp
```

```
spec:
  containers:
  - name: packtapp-container
    image: busybox:latest
    command: ['sh', '-c', 'echo The packtapp is running! &&
sleep 3600']
    initContainers:
  - name: init-pservice
    image: busybox:latest
    command: ['sh', '-c', 'until nslookup packtservice; do echo
waiting for packtservice; sleep 2; done;']
```

Use the following command to deploy the yaml manifest:

kubectl apply -f packt-pod.yaml
Use the following command to see if the pod is up and running:
kubectl get podpackt

Scenario 5

SSH to worker-0, and create a nginx pod and then a busybox container in the same pod.

Create a yaml definition called packt-pod.yaml shown as follows:

```
apiVersion: v1
kind: Pod
metadata:
  name: pactk-multi-pod
  labels:
      app: multi-app
spec:
  containers:
  - name: nginx
    image: nginx
    ports:
    - containerPort: 80
  - name: busybox-sidecar
    image: busybox
    command: ['sh', '-c', 'while true; do sleep 3600; done;']
```

Use the following command to deploy the yaml manifest:

```
kubectl apply -f packt-pod.yaml
Use the following command to see if the pod is up and running:
kubectl get pod pactk-multi-pod
```

Chapter 5 – Demystifying Kubernetes Storage

You have two virtual machines: master-0 and worker-0. Please complete the following mock scenarios.

Scenario 1

Create a new PV called packt-data-pv with a storage of 2GB, and two persistent volume claims (PVCs) requiring 1GB local storage each.

Create a yaml definition called packt-data-pv.yaml for persistent volume as the following:

```
apiVersion: v1
kind: PersistentVolume
metadata:
  name: packt-data-pv
spec:
  storageClassName: local-storage
  capacity:
    storage: 2Gi
  accessModes:
    - ReadWriteOnce
```

Use the following command to deploy the yaml manifest:

```
kubectl apply -f packt-data-pv.yaml
```

Create a yaml definition called packt-data-pvc1.yaml for persistent volume claim as the following:

```
apiVersion: v1
 kind: PersistentVolumeClaim
 metadata:
   name: packt-data-pvc1
 spec:
   storageClassName: local-storage
```

```
accessModes:
    - ReadWriteOnce
resources:
  requests:
    storage: 1Gi
```

Create a yaml definition called `packt-data-pvc2.yaml` for persistent volume claim as the following:

```
apiVersion: v1
 kind: PersistentVolumeClaim
 metadata:
   name: packt-data-pvc2
 spec:
   storageClassName: local-storage
   accessModes:
       - ReadWriteOnce
   resources:
     requests:
         storage: 1Gi
```

Use the following command to deploy the yaml manifest:

```
kubectl apply -f packt-data-pv1.yaml,packt-data-pv2.yaml
```

Scenario 2

Provision a new pod called `packt-storage-pod`, and assign an available PV to this pod.

Create a yaml definition called `packt-data-pod.yaml` shown as follows:

```
apiVersion: v1
 kind: Pod
 metadata:
   name: packt-data-pod
 spec:
   containers:

     - name: busybox
         image: busybox
```

```
        command: ["/bin/sh", "-c","while true; do sleep
3600;  done"]
        volumeMounts:
        - name: temp-data
          mountPath: /tmp/data

    volumes:

      - name: temp-data
        persistentVolumeClaim:
          claimName: packt-data-pv1
      restartPolicy: Always
```

Use the following command to deploy the yaml manifest:

```
kubectl apply -f packt-data-pod.yaml
```

Use the following command to see if the pod is up and running:

```
kubectl get pod packt-data-pod
```

Chapter 6 – Securing Kubernetes

You have two virtual machines: master-0 and worker-0, please complete the following mock scenarios.

Scenario 1

Create a new service account named packt-sa in a new namespace called packt-ns.

Use the following command to create a new service account in the targeting namespace:

```
kubectl create sa packt-sa -n packt-ns
```

Scenario 2

Create a Role named packt-role and bind it with the RoleBinding packt-rolebinding. Map the packt-sa service account with list and get permissions.

Use the following command to create a cluster role in the targeting namespace:

```
kubectl create role packt-role --verb=get --verb=list
--resource=pods --namespace=packt-ns
```

Use the following command to create a Role binding in the targeting namespace:

```
kubectl create rolebinding packt-pods-binding --role=packt-role
--user=packt-user -- namespace=packt-ns
```

To achieve the same result, you can create a yamldefinition called `packt-role.yaml`:

```
apiVersion: rbac.authorization.k8s.io/v1

kind: Role

metadata:

  namespace: packt-ns

  name: packt-clusterrole

rules:

- apiGroups: [""]

  resources: ["pods"]

  verbs: ["get", "list"]
```

Create another yaml definition called `packt-pods-binding.yaml`:

```
apiVersion: rbac.authorization.k8s.io/v1

kind: RoleBinding

metadata:

  name: packt-pods-binding

  namespace: packt-ns
```

```
subjects:

- kind: User

  apiGroup: rbac.authorization.k8s.io

  name:packt-user

roleRef:

  kind: Role

  name: packt-role

  apiGroup: rbac.authorization.k8s.io
```

Use the following command to deploy the yaml manifest:

```
kubectl apply -f packt-role.yaml,packt-pods-binding.yaml
```

Verify the Role using the following command:

```
kubectl get roles -n packt-ns
```

Verify the rolebindings by using the following command:

```
kubectl get rolebindings -n packt-ns
```

Scenario 3

Create a new pod named packt-pod with the image busybox:1.28 in the namespace packt-ns. Expose port 80. Then assign the service account packt-sa to the pod.

Use the following command to create a deployment:

```
kubectl create deployment packtbusybox --image=busybox:1.28 -n
packt-ns –port 80
```

Export the deployment information in yaml specification form:

```
kubectl describe deployment packtbusybox -n packt-ns -o yaml >
packt-busybox.yaml
```

Edit the yaml specification to reference the service account:

```
apiVersion: v1
kind: Deployment
metadata:
  name: packtbusybox
  namespace : packt-ns
spec:
  containers:
  - image: busybox
    name: packtbusybox
    volumeMounts:
    - mountPath: /var/run/secrets/tokens
      name: vault-token
  serviceAccountName: packt-sa
  volumes:
  - name: vault-token
    projected:
      sources:
      - serviceAccountToken:
          path: vault-token
          expirationSeconds: 7200
          audience: vault
```

Check out the *Implementing Kubernetes RBAC* section in *Chapter 6, Securing Kubernetes* to get further information about how to implement RBAC.

Chapter 7 – Demystifying Kubernetes networking

You have two virtual machines: master-0 and worker-0. Please complete the following mock scenarios.

Scenario 1

Deploy a new deployment nginx with the latest image of nginx for 2 replicas, in a namespace called packt-app. The container is exposed on port 80. Create a service type ClusterIP within the same namespace. Deploy a sandbox-nginx pod and make a call using curl to verify the connectivity to the nginx service.

Use the following command to create nginx deployment in the targeting namespace:

```
kubectl create deployment nginx --image=nginx --replicas=2 -n
packt-app
```

Use the following command to expose nginx deployment with a ClusterIP service in the targeting namespace:

```
kubectl expose deployment nginx --type=ClusterIP --port 8080
--name=packt-svc --target-port 80 -n packt-app
```

Use the following command to get the internal IP:

```
kubectl get nodes -o jsonpath='{.items[*].status.addresses[?(
@.type=="INTERNAL-IP")].address}'
```

Use the following command to get the endpoint:

```
kubectl get svc packt-svc -n packt-app -o wide
```

Use the following command to deploy a sandbox-nginx pod in the targeting namespace using your endpoint:

```
kubectl run -it sandbox-nginx --image=nginx -n packt-app --rm
--restart=Never -- curl -Is http://192.168.xx.x (internal IP
):31400 ( endpoint )
```

Scenario 2

Expose the nginx deployment with the NodePort service type; the container is exposed on port 80. Use the test-nginx pod to make a call using curl to verify the connectivity to the nginx service.

Use the following command to create nginx deployment in the targeting namespace:

```
kubectl expose deployment nginx --type=NodePort --port 8080
--name=packt-svc --target-port 80 -n packt-app
```

Use the following command to get the internal IP:

```
kubectl get nodes -o jsonpath='{.items[*].status.addresses[?(
@.type=="INTERNAL-IP")].address}'
```

Use the following command to get the endpoint:

```
kubectl get svc packt-svc -n packt-app -o wide
```

Use the following command to deploy a test-nginx pod in the targeting namespace using your endpoint:

```
kubectl run -it test-nginx --image=nginx -n packt-app --rm
--restart=Never -- curl -Is http://192.168.xx.x (internal IP
):31400 ( endpoint )
```

Scenario 3

Make a call using wget or curl from the machine within the same network with that node, to verify the connectivity with the nginx NodePort service through the correct port.

Call from worker-2 using the following command:

```
curl -Is http://192.168.xx.x (internal IP of the worker 2
):31400 ( the port of that node  )
```

Alternatively, we can use wget as the following command:

```
wget http://192.168.xx.x (internal IP of the worker 2 ):31400 (
the port of that node  )
```

Scenario 4

Use the sandbox-nginx pod to nslookup the IP address of nginx NodePort service. See what is returned.

Use the following command:

```
kubectl run -it sandbox-nginx --image=busybox:latest
kubect exec sandbox-nginx -- nslookup <ip address of nginx
Nodeport>
```

Scenario 5

Use the sandbox-nginx pod to nslookup the DNS domain hostname of nginx NodePort service. See what is returned.

Use the following command:

```
kubectl run -it sandbox-nginx --image=busybox:latest
kubect exec sandbox-nginx -- nslookup <hostname of nginx
Nodeport>
```

Scenario 6

Use the sandbox-nginx pod to `nslookup` the DNS domain hostname of nginx pod. See what is returned.

Use the following command:

```
kubectl run -it sandbox-nginx --image=busybox:latest
kubect exec sandbox-nginx -- nslookup x-1-0-9(pod ip address).
pack-app.pod.cluster.local
```

Chapter 8 – Monitoring and logging Kubernetes Clusters and Applications

You have two virtual machines: `master-0` and `worker-0`. Please complete the following mock scenarios.

Scenario 1

List all the available pods in your current cluster and find what the most CPU-consuming pods are. Write the name to the `max-cpu.txt` file.

Use the following command:

```
kubectl top pod -- all-namespaces --sort-by=cpu > max-cpu.txt
```

Index

Packt.com

Subscribe to our online digital library for full access to over 7,000 books and videos, as well as industry leading tools to help you plan your personal development and advance your career. For more information, please visit our website.

Why subscribe?

- Spend less time learning and more time coding with practical eBooks and Videos from over 4,000 industry professionals
- Improve your learning with Skill Plans built especially for you
- Get a free eBook or video every month
- Fully searchable for easy access to vital information
- Copy and paste, print, and bookmark content

Did you know that Packt offers eBook versions of every book published, with PDF and ePub files available? You can upgrade to the eBook version at packt.com and as a print book customer, you are entitled to a discount on the eBook copy. Get in touch with us at customercare@packtpub.com for more details.

At www.packt.com, you can also read a collection of free technical articles, sign up for a range of free newsletters, and receive exclusive discounts and offers on Packt books and eBooks.

Other Books You May Enjoy

If you enjoyed this book, you may be interested in these other books by Packt:

End-to-End Automation with Kubernetes and Crossplane

Arun Ramakani

ISBN: 9781801811545

- Understand the context of Kubernetes-based infrastructure automation
- Get to grips with Crossplane concepts with the help of practical examples
- Extend Crossplane to build a modern infrastructure automation platform
- Use the right configuration management tools in the Kubernetes environment
- Explore patterns to unify application and infrastructure automation
- Discover top engineering practices for infrastructure platform as a product

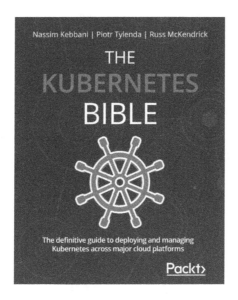

The Kubernetes Bible

Nassim Kebbani, Piotr Tylenda, Russ McKendrick

ISBN: 9781838827694

- Manage containerized applications with Kubernetes
- Understand Kubernetes architecture and the responsibilities of each component
- Set up Kubernetes on Amazon Elastic Kubernetes Service, Google Kubernetes Engine, and Microsoft Azure Kubernetes Service
- Deploy cloud applications such as Prometheus and Elasticsearch using Helm charts
- Discover advanced techniques for Pod scheduling and auto-scaling the cluster
- Understand possible approaches to traffic routing in Kubernetes

Packt is searching for authors like you

If you're interested in becoming an author for Packt, please visit authors.packtpub.com and apply today. We have worked with thousands of developers and tech professionals, just like you, to help them share their insight with the global tech community. You can make a general application, apply for a specific hot topic that we are recruiting an author for, or submit your own idea.

Share Your Thoughts

Now you've finished *Certified Kubernetes Administrator (CKA) Exam Guide* , we'd love to hear your thoughts! Scan the QR code below to go straight to the Amazon review page for this book and share your feedback or leave a review on the site that you purchased it from.

https://packt.link/r/1803238267

Your review is important to us and the tech community and will help us make sure we're delivering excellent quality content.